400 Science Facts for Kids

Fun Science book for Kids to Learn About Earth Science, Planets, Chemistry, Physics, Biology, Plants and More

JAKE BLANTON

Copyright © 2023 by JAKE BLANTON

All rights reserved. No part of this publication may be reproduced, distributed, or transmitted in any form or by any means, including photocopying, recording, or other electronic or mechanical methods, without the prior written permission of the publisher.

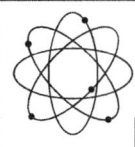   

CONTENTS

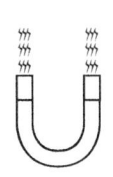 

Earth Science

Planets

Chemistry

Physics

Biology

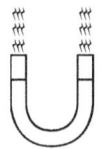

Plants

Famous Scientist

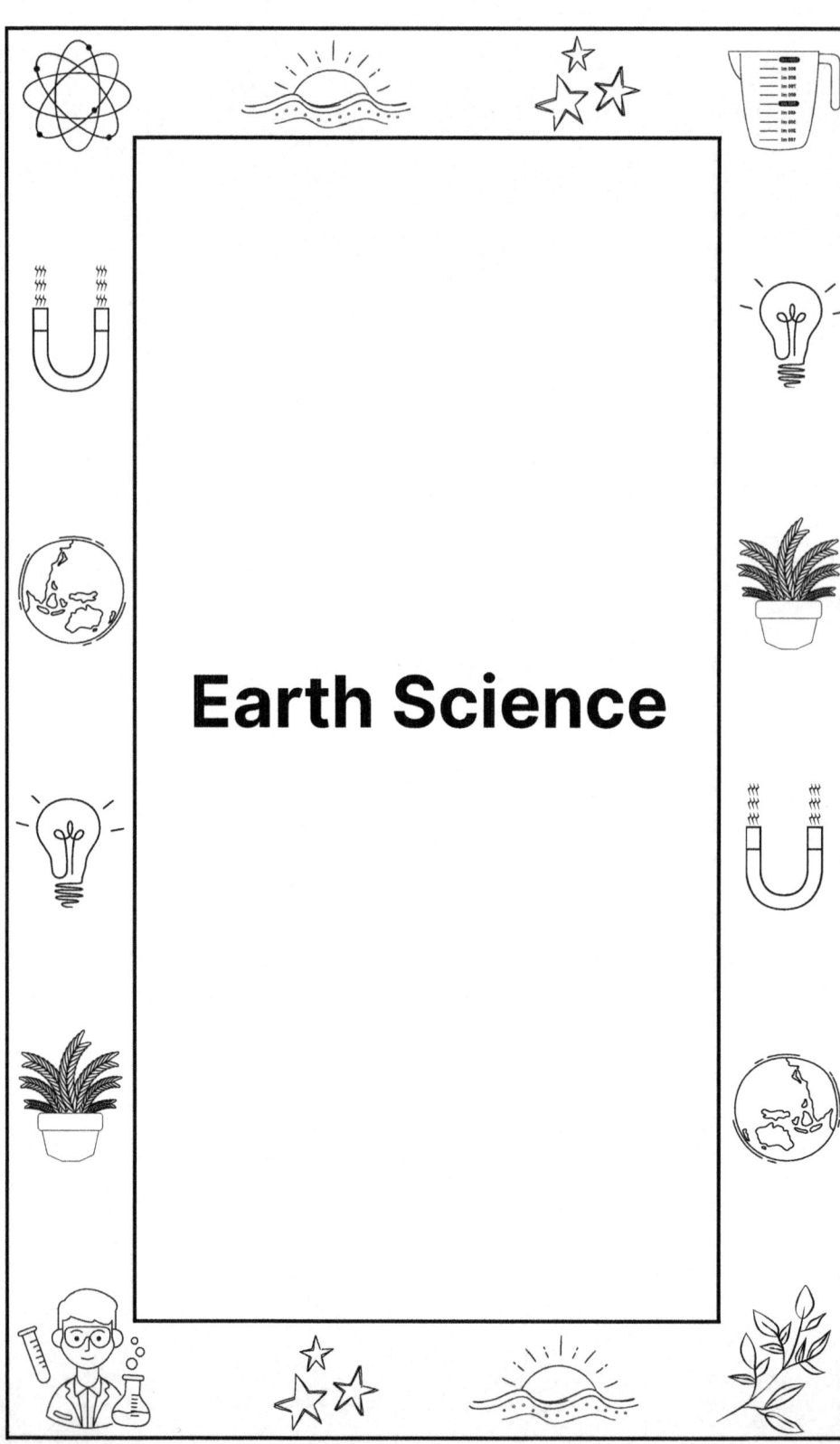

Earth Science

Earth is 4.5 billion years old.

Earth is the only planet known to have liquid water on its surface.

Earth's oceans contain 97% of the planet's water.

Earth's gravity is what keeps all of the planet's mass in place.

Earth's rotation on its axis causes the change of seasons.

Earth's atmosphere helps regulate the planet's temperature.

Earth's average distance from the sun is about 93 million miles (149.6 million kilometers).

Earth's oceans are divided into five main regions: the Atlantic, Pacific, Indian, Southern, and Arctic.

Earth rotates at around 1000 miles an hour.

Earth has a diameter of about 12,742 kilometers (7,917 miles).

Earth's average distance from the sun is about 8 light-minutes.

Earth's gravity is weaker at higher altitudes, which is why objects weigh less on mountains.

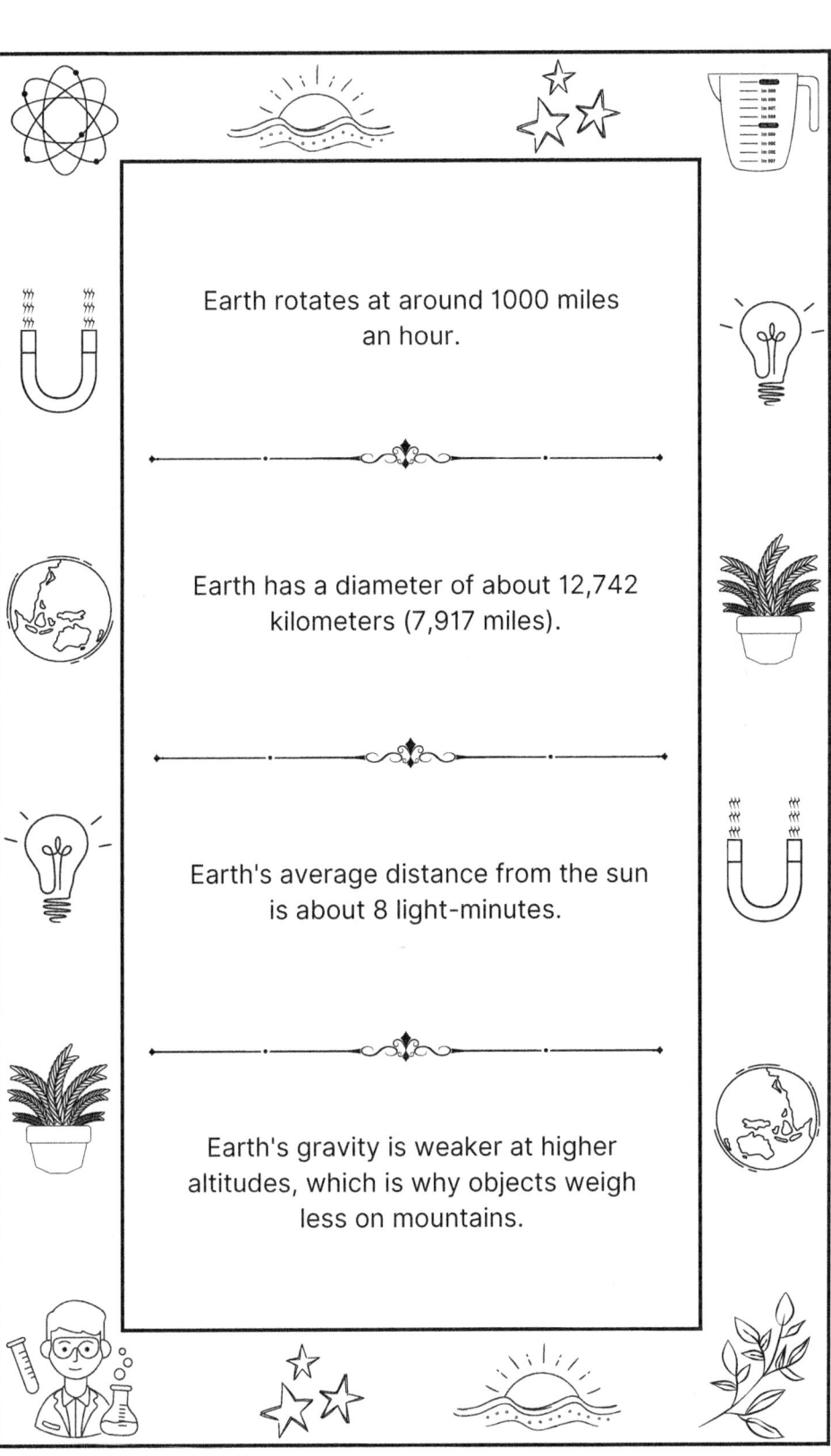

Earth is the third planet from the Sun.

Earth's average surface temperature is about 15°C (59°F).

Earth's atmosphere contains clouds, which are formed when moist air rises and cools.

Earth's oceans are home to many types of coral reefs.

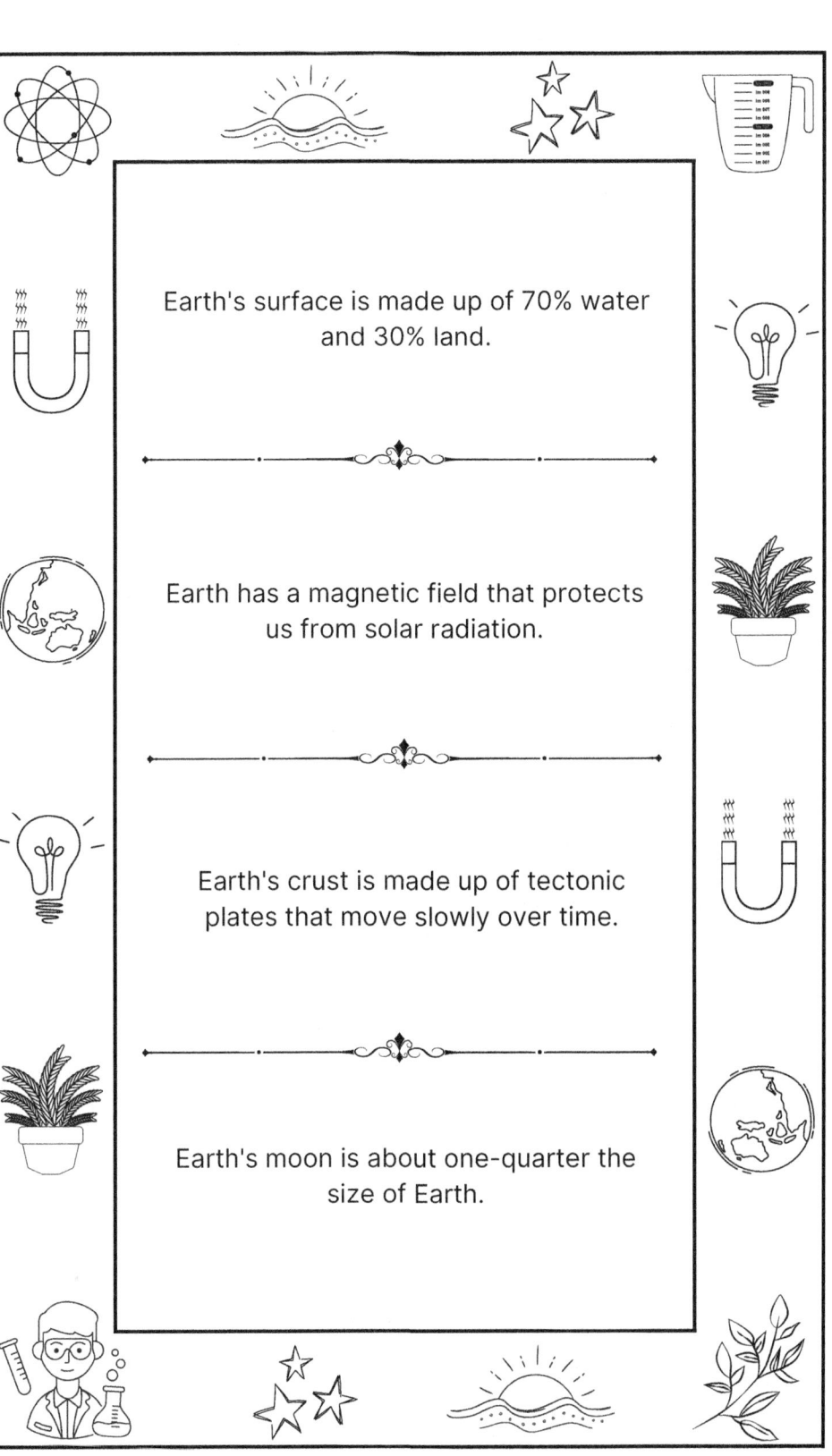

Earth's surface is made up of 70% water and 30% land.

Earth has a magnetic field that protects us from solar radiation.

Earth's crust is made up of tectonic plates that move slowly over time.

Earth's moon is about one-quarter the size of Earth.

Earth's atmosphere contains ozone, which helps protect us from the sun's harmful UV rays.

Earth's temperature is regulated by the amount of greenhouse gases in the atmosphere.

Earth's oceans are home to a diverse array of marine life.

Earth's atmosphere helps to preserve our planet's water supply.

Earth's atmosphere is made up of 78% nitrogen, 21% oxygen, and trace amounts of other gases.

Earth's atmosphere helps to regulate the planet's temperature by trapping some of the sun's heat.

Earth's average surface temperature has increased by about 1°C (1.8°F) since the industrial revolution.

Earth's oceans play a vital role in regulating the planet's climate.

Earth's oceans are home to many types of marine mammals, such as whales, dolphins, and seals.

Earth's oceans contain a variety of dissolved minerals, including salt, calcium, and magnesium.

Earth's atmosphere contains trace amounts of carbon monoxide, which is a poisonous gas produced by the burning of fossil fuels.

The Earth's path around the Sun is called its orbit.

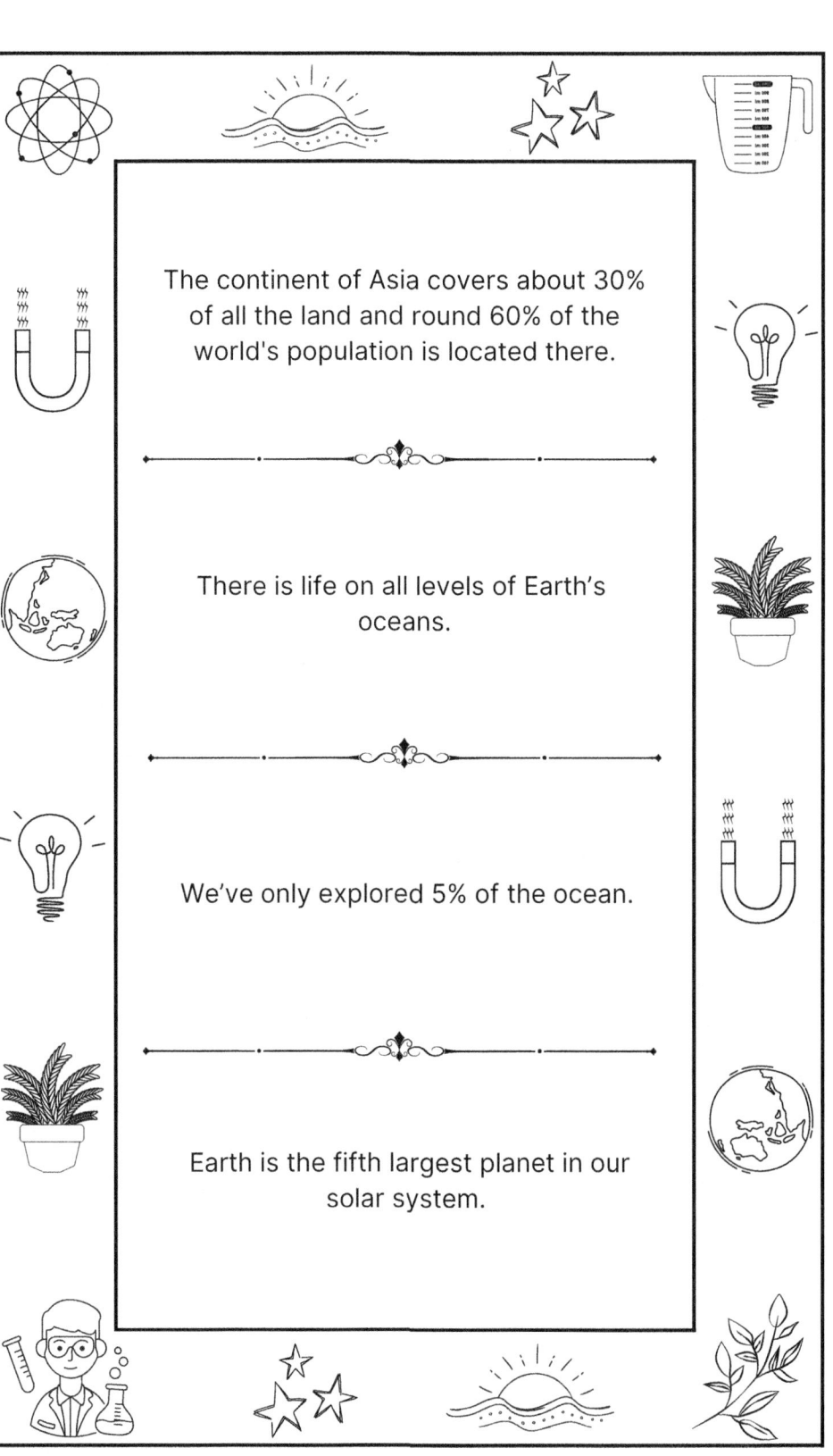

The continent of Asia covers about 30% of all the land and round 60% of the world's population is located there.

There is life on all levels of Earth's oceans.

We've only explored 5% of the ocean.

Earth is the fifth largest planet in our solar system.

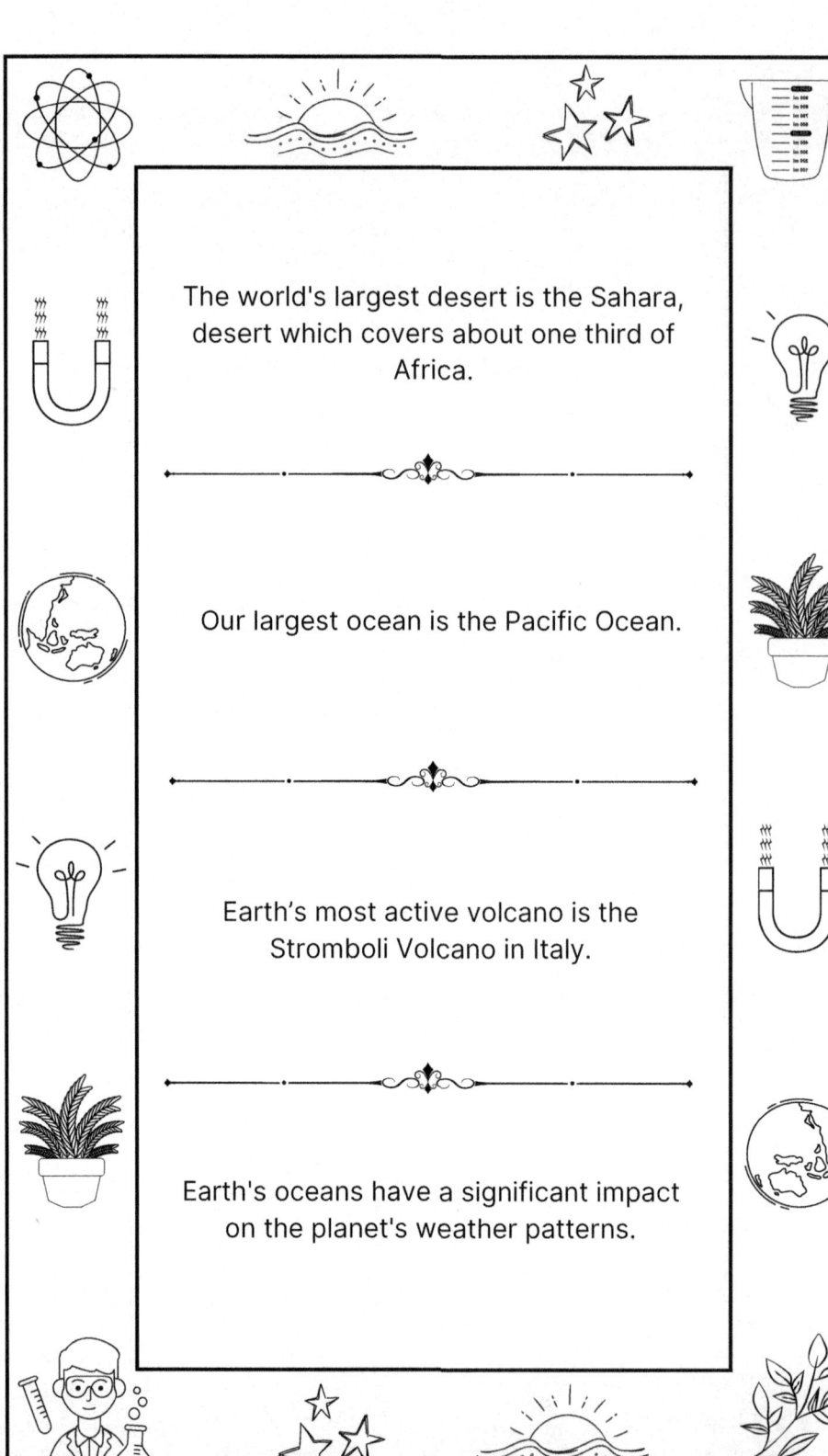

The world's largest desert is the Sahara, desert which covers about one third of Africa.

Our largest ocean is the Pacific Ocean.

Earth's most active volcano is the Stromboli Volcano in Italy.

Earth's oceans have a significant impact on the planet's weather patterns.

Earth's atmosphere is made up of different layers, including the troposphere, stratosphere, and mesosphere.

Earth is the only planet in our solar system known to support life.

We live on the thin outer layer of the Earth that is Crust.

Earth has a diverse range of habitats such as mountains, forests, deserts and oceans.

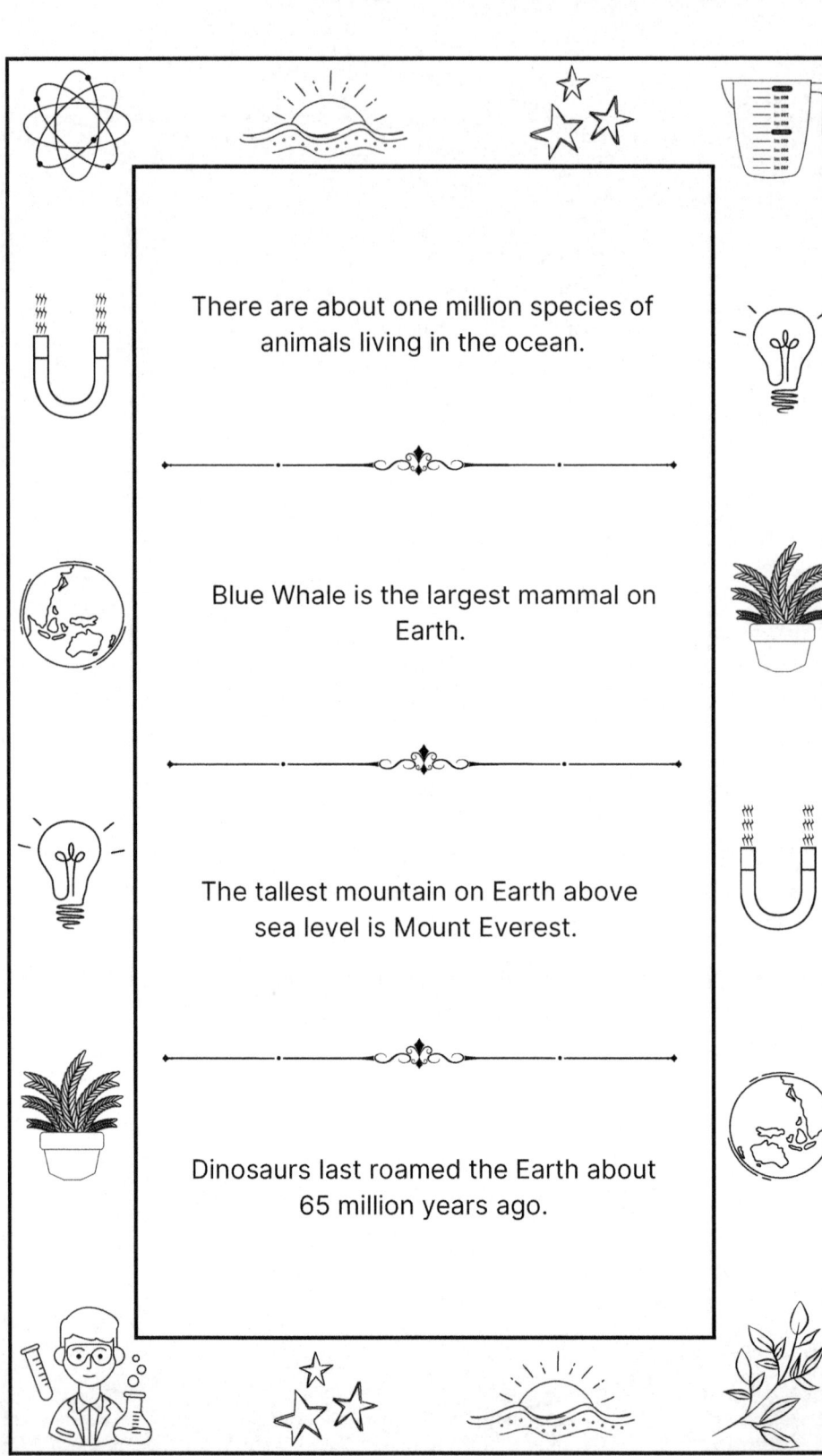

There are about one million species of animals living in the ocean.

Blue Whale is the largest mammal on Earth.

The tallest mountain on Earth above sea level is Mount Everest.

Dinosaurs last roamed the Earth about 65 million years ago.

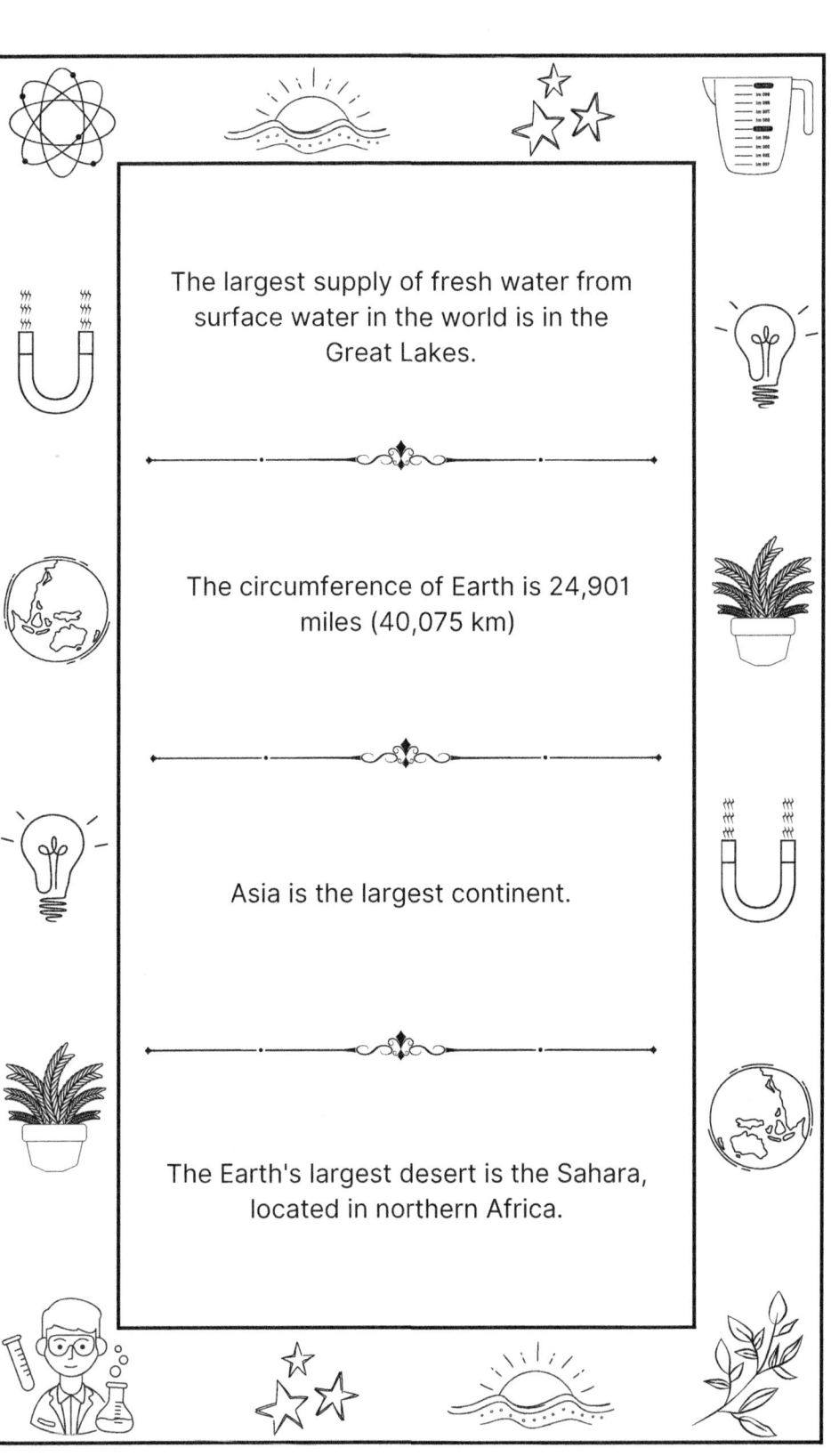

The largest supply of fresh water from surface water in the world is in the Great Lakes.

The circumference of Earth is 24,901 miles (40,075 km)

Asia is the largest continent.

The Earth's largest desert is the Sahara, located in northern Africa.

Africa is the second-largest continent.

Death Valley in California is commonly known as the hottest place in the world.

Some hikers have summited Everest without oxygen.

The most snowfall per year happens in Japan.

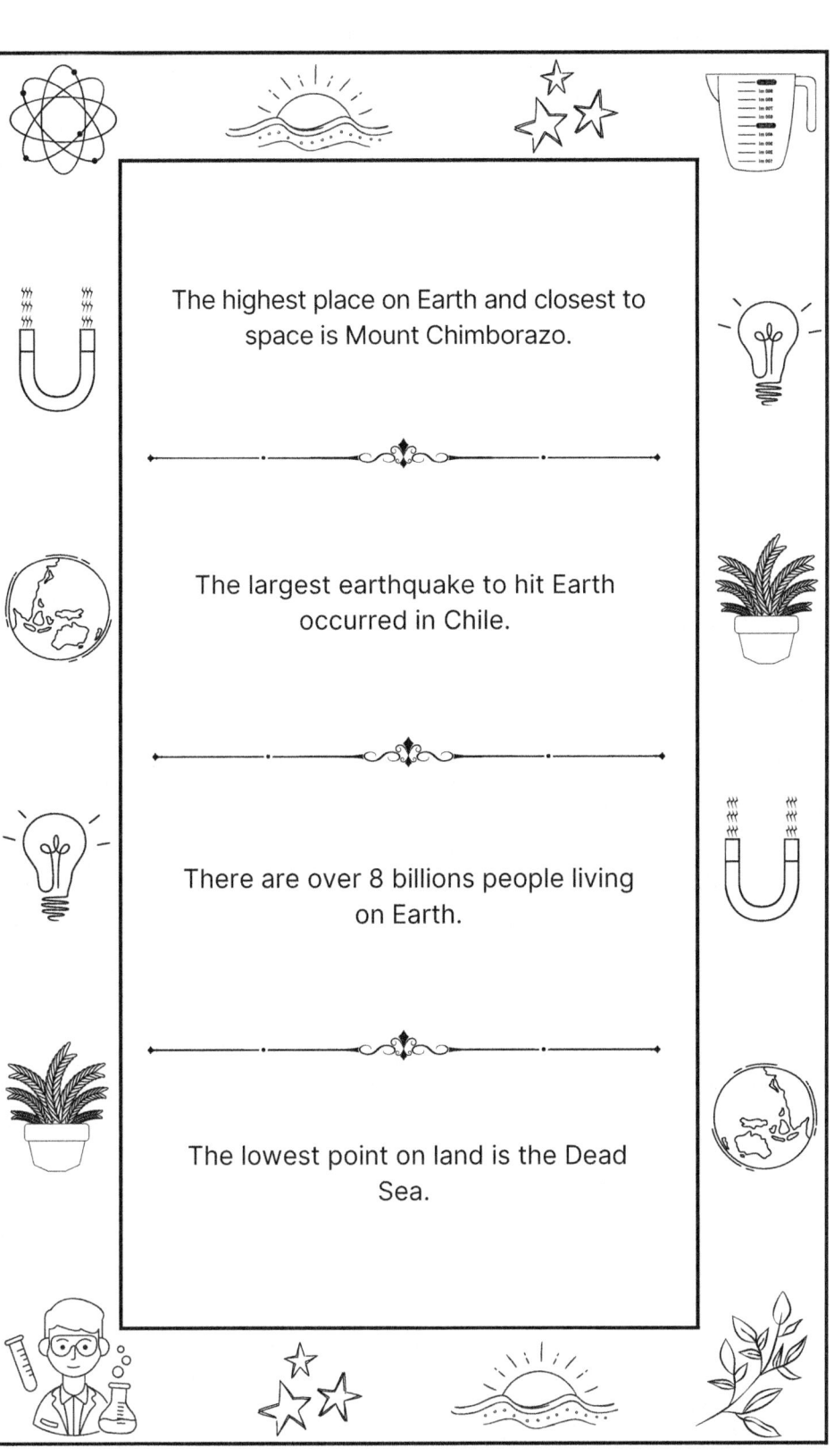

The highest place on Earth and closest to space is Mount Chimborazo.

The largest earthquake to hit Earth occurred in Chile.

There are over 8 billions people living on Earth.

The lowest point on land is the Dead Sea.

The deepest point on the ocean floor is 36,200 feet below sea level.

Antarctica is technically a desert.

About 90% of Earth's freshwater is locked in ice.

The Earth was covered with giant mushrooms before trees.

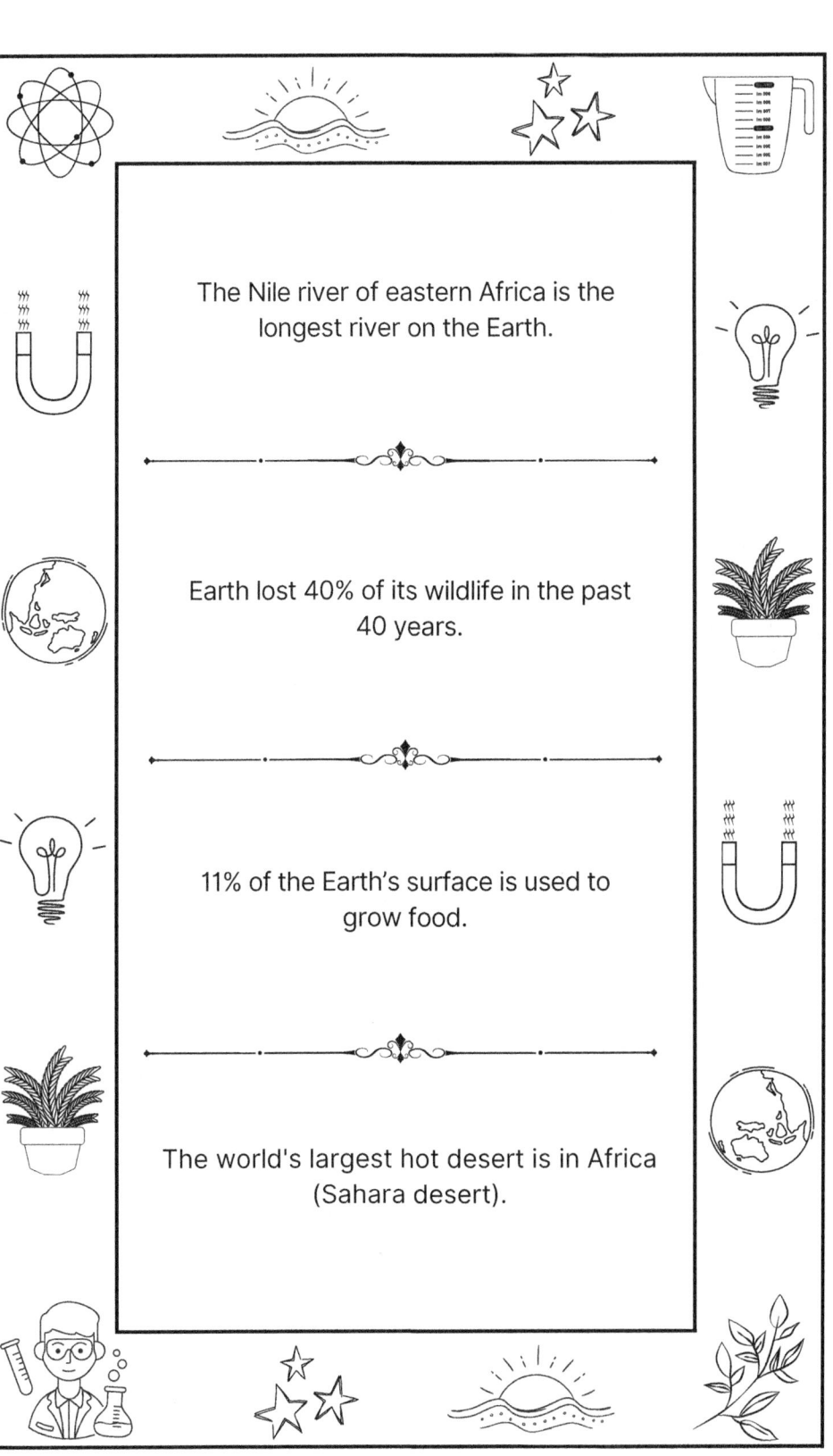

The Nile river of eastern Africa is the longest river on the Earth.

Earth lost 40% of its wildlife in the past 40 years.

11% of the Earth's surface is used to grow food.

The world's largest hot desert is in Africa (Sahara desert).

Planets

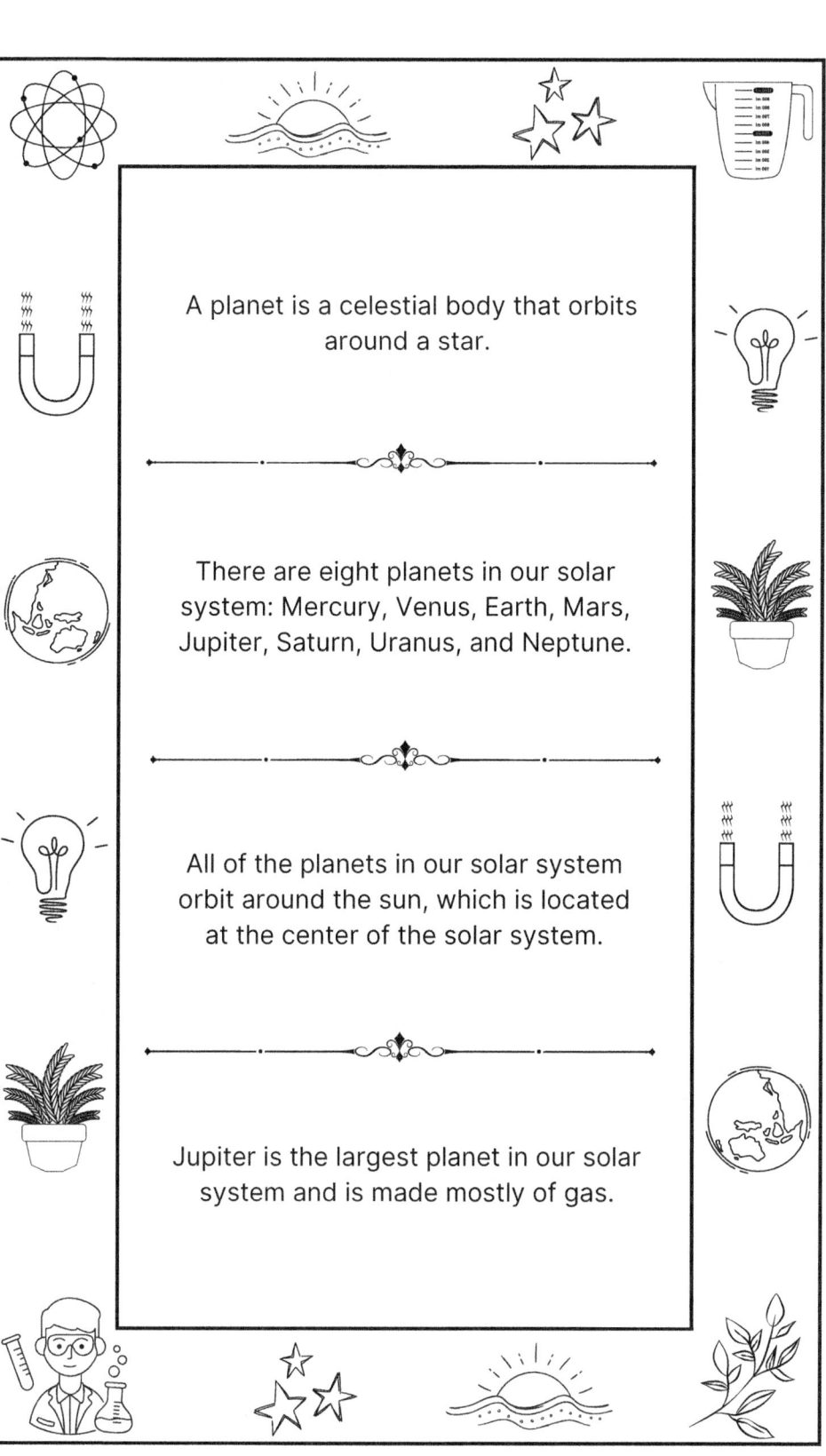

A planet is a celestial body that orbits around a star.

There are eight planets in our solar system: Mercury, Venus, Earth, Mars, Jupiter, Saturn, Uranus, and Neptune.

All of the planets in our solar system orbit around the sun, which is located at the center of the solar system.

Jupiter is the largest planet in our solar system and is made mostly of gas.

Saturn is the second largest planet in our solar system and is known for its rings.

Mercury is the smallest planet in our solar system and is also the closest to the sun.

Some planets, like Earth, have a thick atmosphere that helps to protect them from space debris.

Planets are made up of different materials, including rock, gas, and ice.

Planets can have different shapes, including round, oval, and irregular.

The temperature on different planets can vary greatly, from boiling hot on Venus to freezing cold on Uranus.

Planets can have different numbers of seasons, depending on their distance from their star and their tilt.

The surface of a planet can be affected by tectonic activity, such as earthquakes and volcanoes.

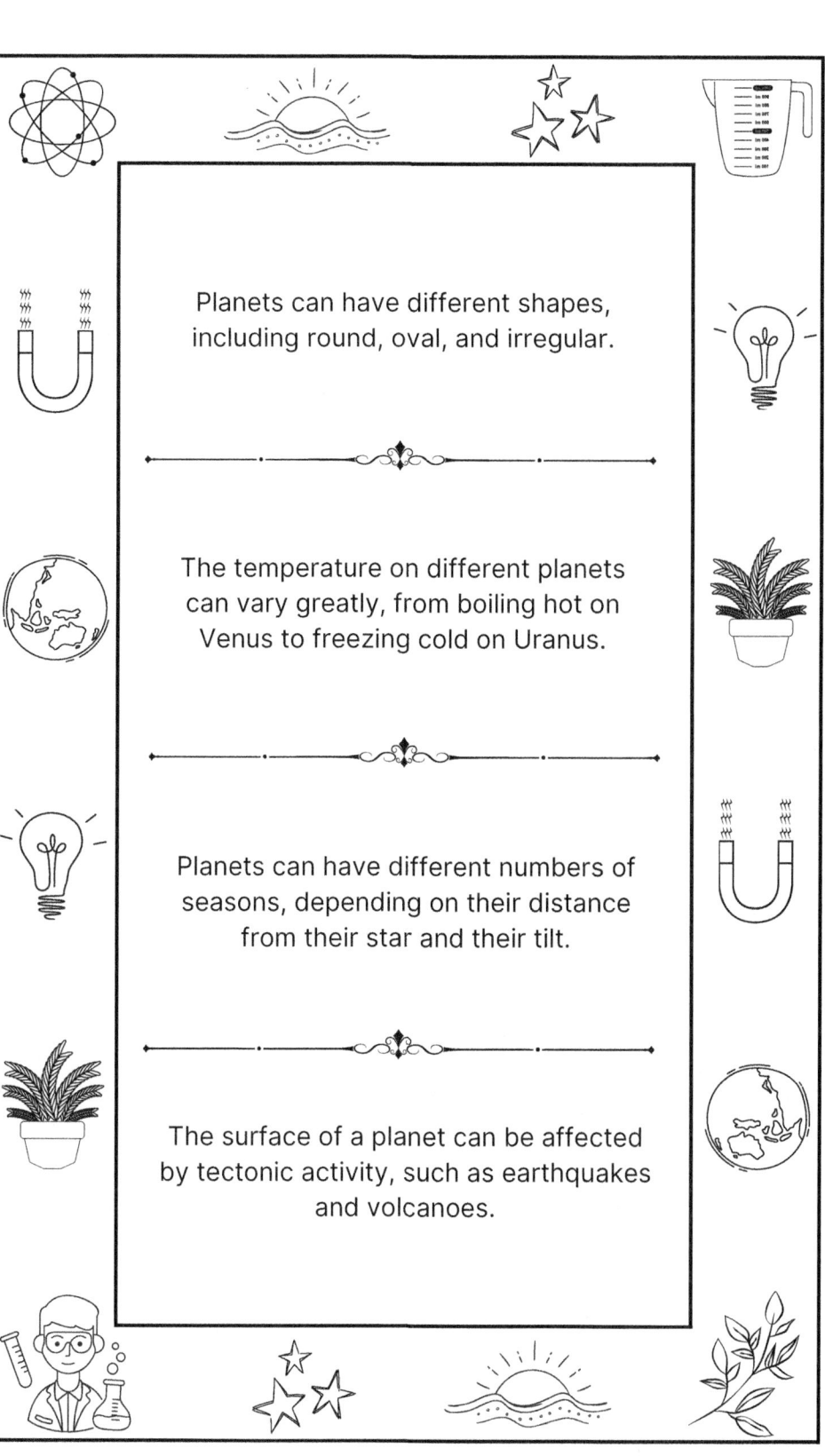

Mercury has no moons or rings.

It takes the Moon about 27 days to complete one orbit around Earth.

Uranus is a gas giant planet and is the third largest planet in our Solar System.

The Moon is the second brightest object in our sky after the Sun.

The Sun's diameter is 1,392,000 kilometres (865,000 miles).

Life on Earth depends on heat and light from the Sun.

Venus is about the same size and weight as Earth

Mars is the fourth planet from the Sun, between Earth and Jupiter.

Some planets, like Mars, have a thin atmosphere that is not able to protect them from space debris.

Uranus has 27 known moons, the largest of which is called Titania.

Some planets, like Jupiter, have strong winds and storms that can last for years.

Neptune is the eighth, and the farthest, planet from the Sun.

Comets are small chunks of ice and dust that orbit the Sun in the solar system.

The mass of a planet can affect its gravitational pull and its ability to hold onto its atmosphere.

The whole of Mars is as cold as the South Pole.

There are billions of comets in the solar system.

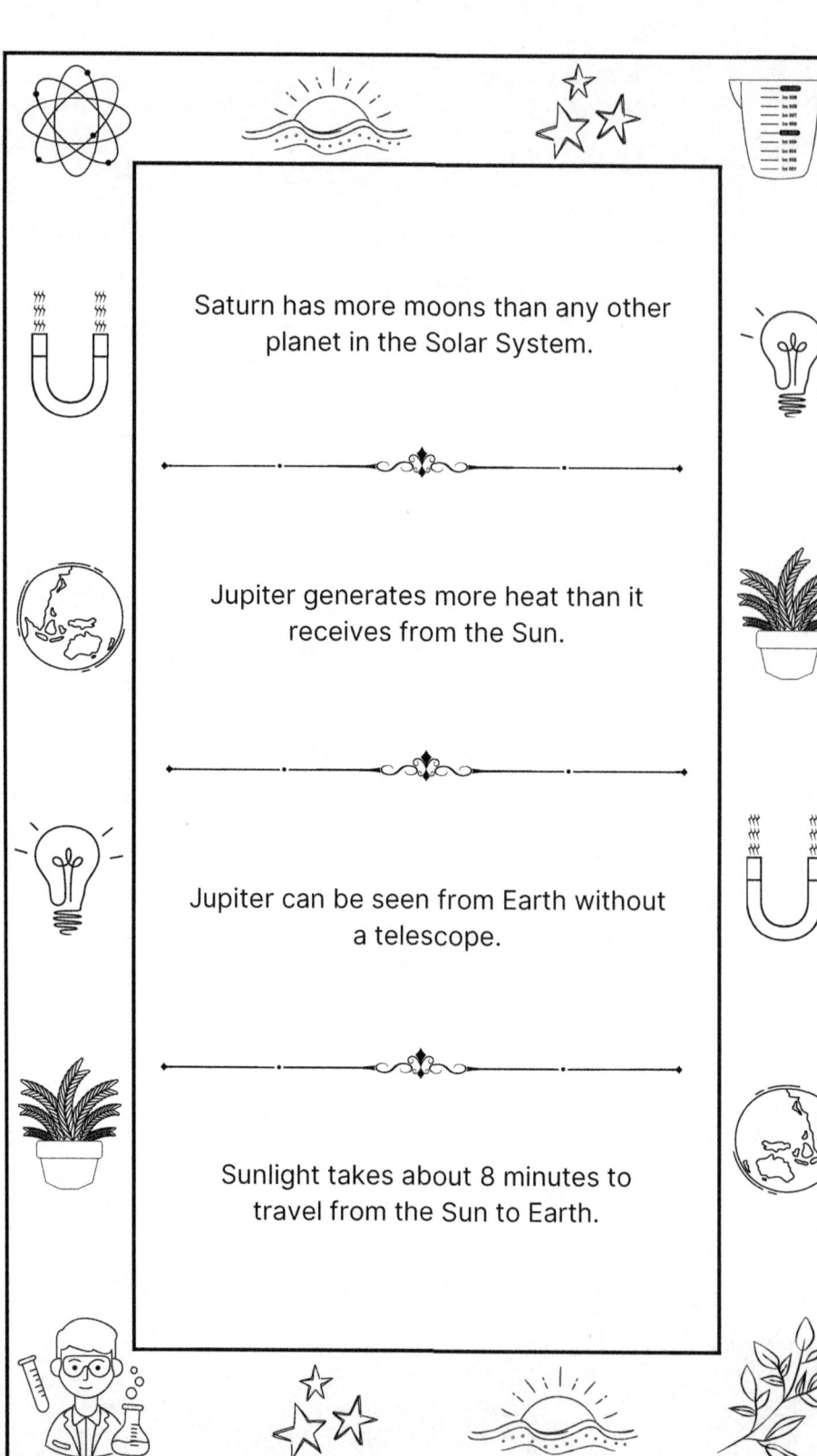

Saturn has more moons than any other planet in the Solar System.

Jupiter generates more heat than it receives from the Sun.

Jupiter can be seen from Earth without a telescope.

Sunlight takes about 8 minutes to travel from the Sun to Earth.

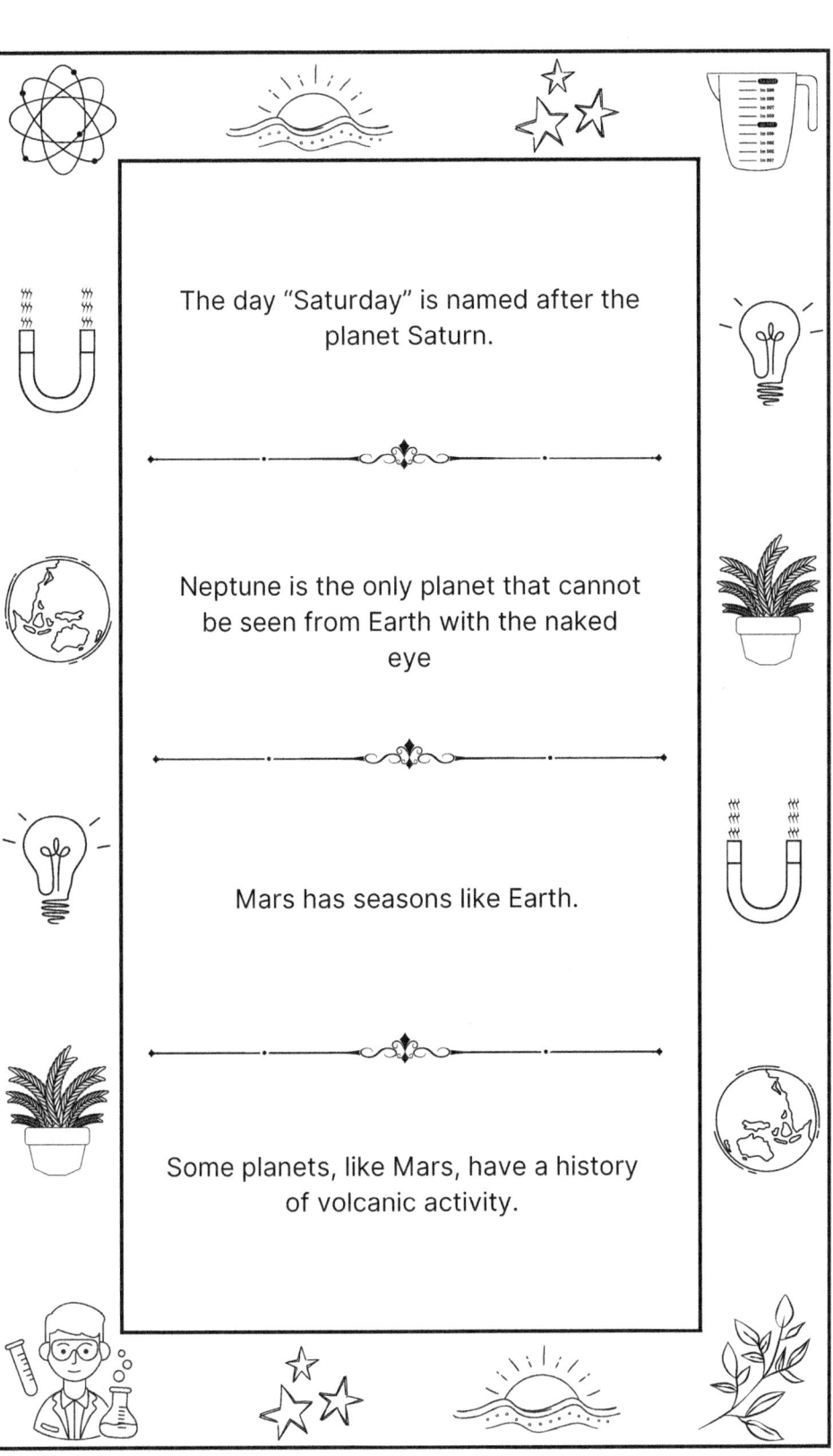

The day "Saturday" is named after the planet Saturn.

Neptune is the only planet that cannot be seen from Earth with the naked eye

---

Mars has seasons like Earth.

Some planets, like Mars, have a history of volcanic activity.

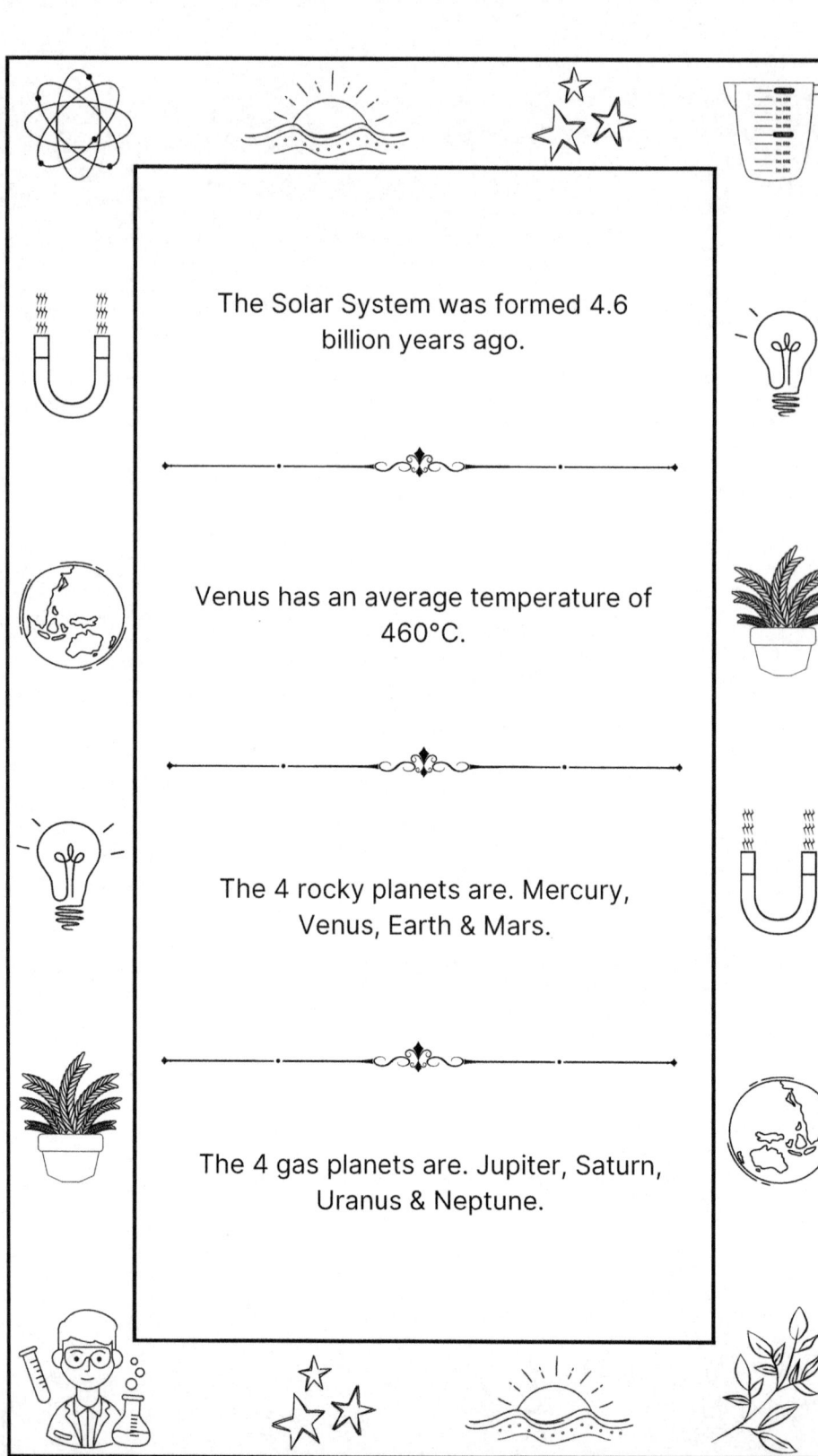

The Solar System was formed 4.6 billion years ago.

Venus has an average temperature of 460°C.

The 4 rocky planets are. Mercury, Venus, Earth & Mars.

The 4 gas planets are. Jupiter, Saturn, Uranus & Neptune.

Mercury has a very thin atmosphere, or exosphere, made up of atoms that have escaped from its surface.

The order of the planets from the sun, starting with the one closest to the sun and moving outward, is: Mercury, Venus, Earth, Mars, Jupiter, Saturn, Uranus, and Neptune.

There may be other planets beyond our solar system that we have not yet discovered.

The Sun is 93 million miles from the Earth.

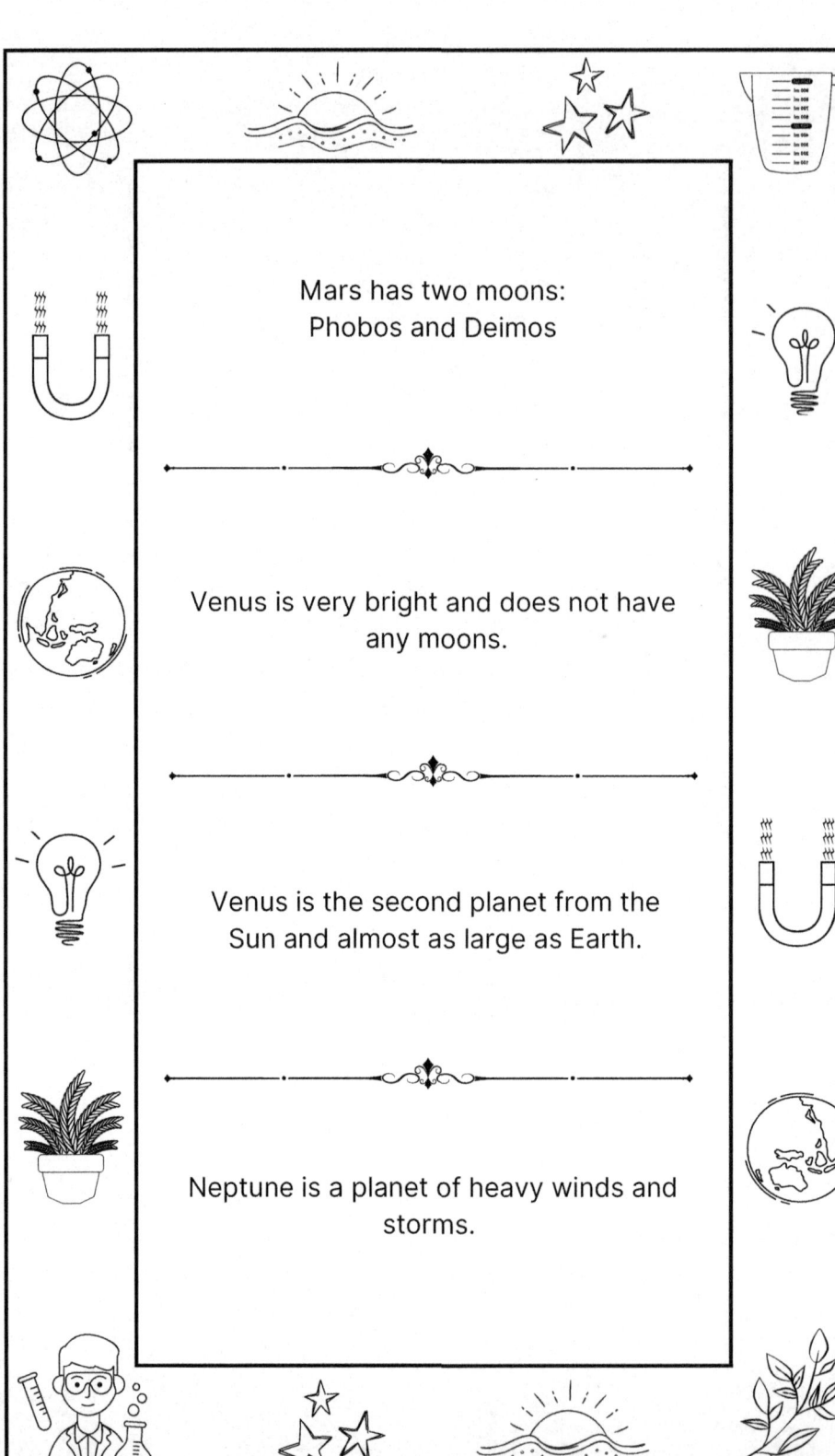

Mars has two moons: Phobos and Deimos

Venus is very bright and does not have any moons.

Venus is the second planet from the Sun and almost as large as Earth.

Neptune is a planet of heavy winds and storms.

The rotation of a planet can affect the length of its day and year.

A spacecraft takes about 3 days to travel from the Earth to the Moon.

Some planets have extreme weather, such as lightning storms on Jupiter and dust storms on Mars.

There are pieces of Mars that have fallen on Earth.

Chemistry

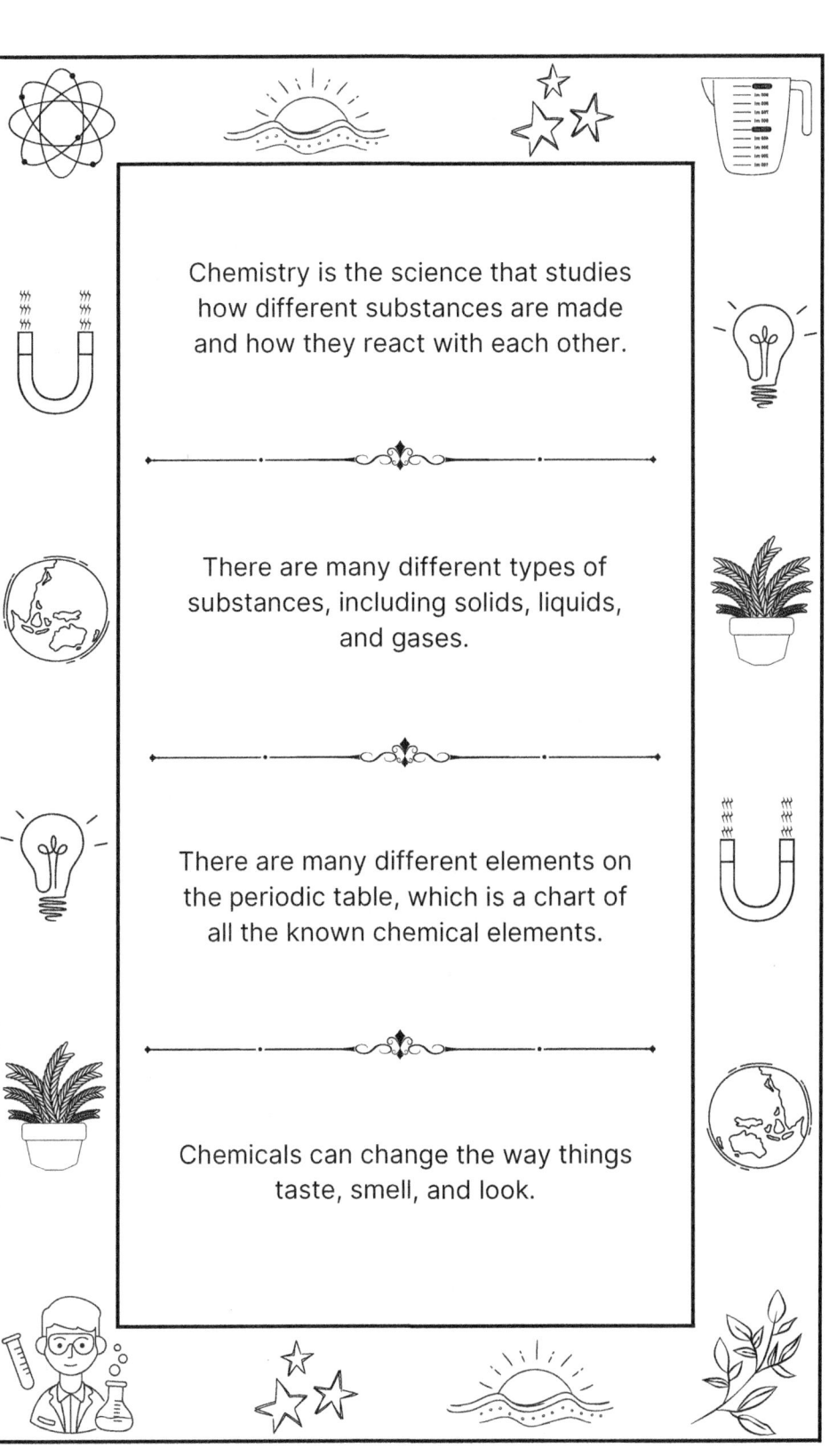

Chemistry is the science that studies how different substances are made and how they react with each other.

There are many different types of substances, including solids, liquids, and gases.

There are many different elements on the periodic table, which is a chart of all the known chemical elements.

Chemicals can change the way things taste, smell, and look.

Each element is made up of atoms, which are the smallest unit of matter.

Atoms are made up of even smaller particles called protons, neutrons, and electrons.

Bromine and Mercury are the only elements that are liquid at room temperature.

When atoms join together, they form molecules.

There are many different types of chemical reactions, including synthesis, decomposition, and exchange reactions.

Some substances dissolve in water, while others do not.

Acids and bases are two special types of chemicals that have different properties.

Gold and copper are the only two non-silvery metals.

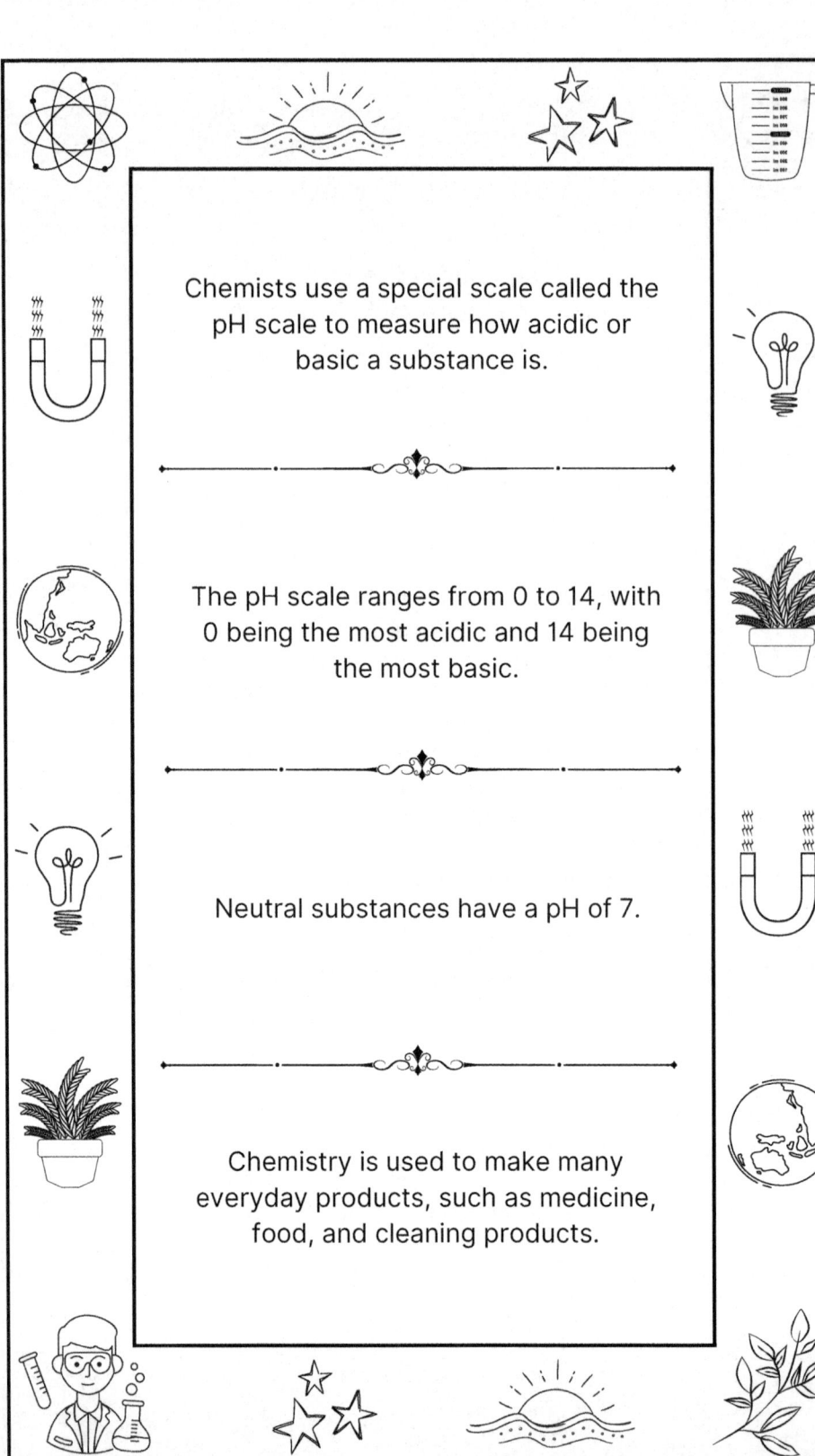

Chemists use a special scale called the pH scale to measure how acidic or basic a substance is.

The pH scale ranges from 0 to 14, with 0 being the most acidic and 14 being the most basic.

Neutral substances have a pH of 7.

Chemistry is used to make many everyday products, such as medicine, food, and cleaning products.

Some substances are conductors, which means they allow electricity to flow through them.

Other substances are insulators, which means they do not allow electricity to flow through them.

Water, for example, is a compound made up of hydrogen and oxygen atoms.

Oxygen is a gas that is essential for life.

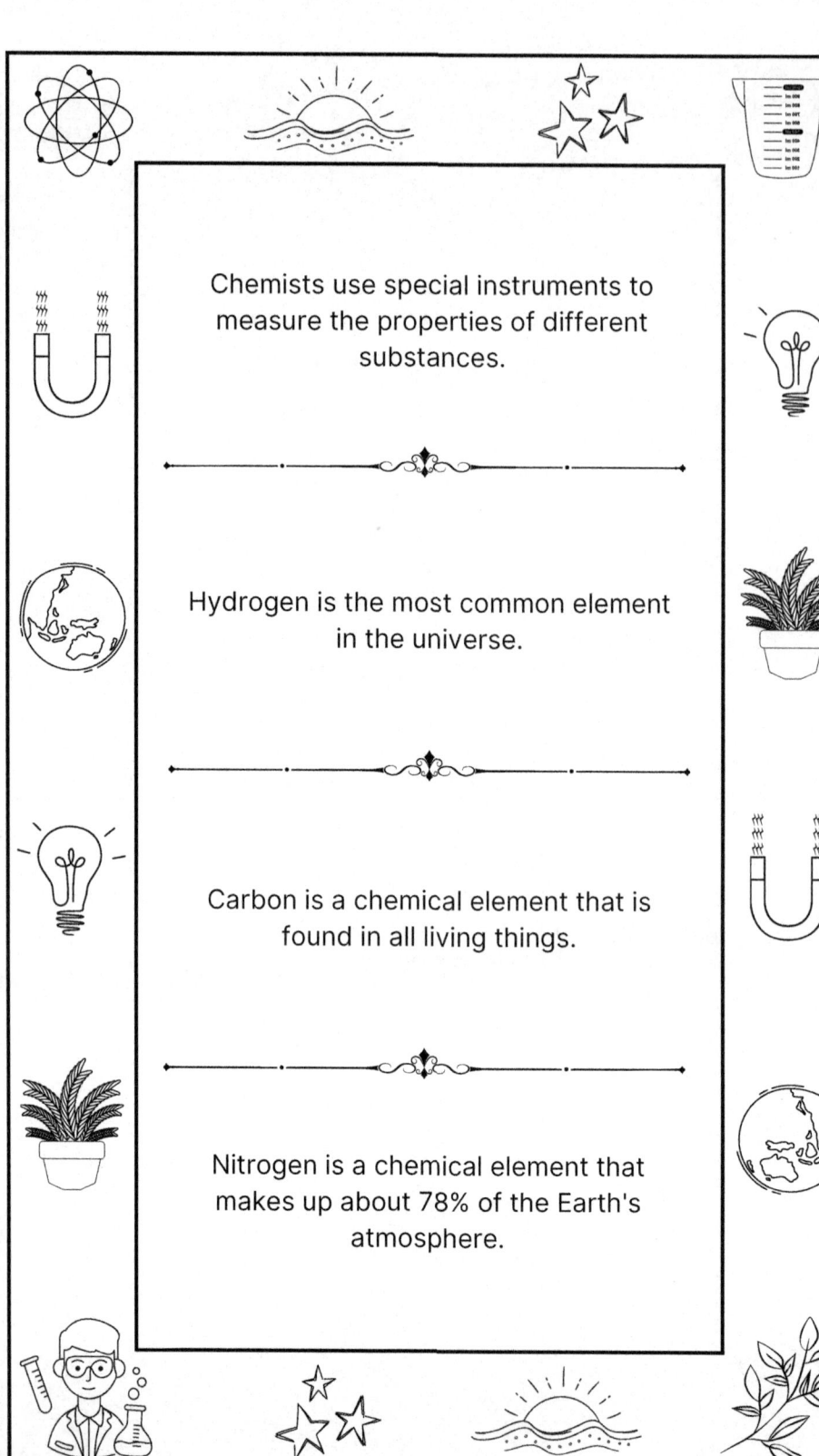

Chemists use special instruments to measure the properties of different substances.

Hydrogen is the most common element in the universe.

Carbon is a chemical element that is found in all living things.

Nitrogen is a chemical element that makes up about 78% of the Earth's atmosphere.

Sodium is a chemical element that is found in table salt.

Chlorine is a chemical element that is used to make bleach.

Potassium is a chemical element that is found in many fruits and vegetables.

Iron is a chemical element that is found in red blood cells.

Calcium is a chemical element that is found in bones and teeth.

Copper is a chemical element that is used to make wires and pipes.

Gold is a chemical element that is prized for its beauty and rarity.

Silver is a chemical element that is used to make jewelry and coins.

Platinum is a chemical element that is used to make jewelry and catalytic converters.

Mercury is a chemical element that is a liquid at room temperature.

Dry ice is the solid form of carbon dioxide.

About half of Earth's oxygen comes from the ocean.

The process of changing a solid into a liquid is called melting.

The process of changing a liquid into a gas is called vaporization.

The process of changing a gas into a liquid is called condensation.

The process of changing a liquid into a solid is called freezing.

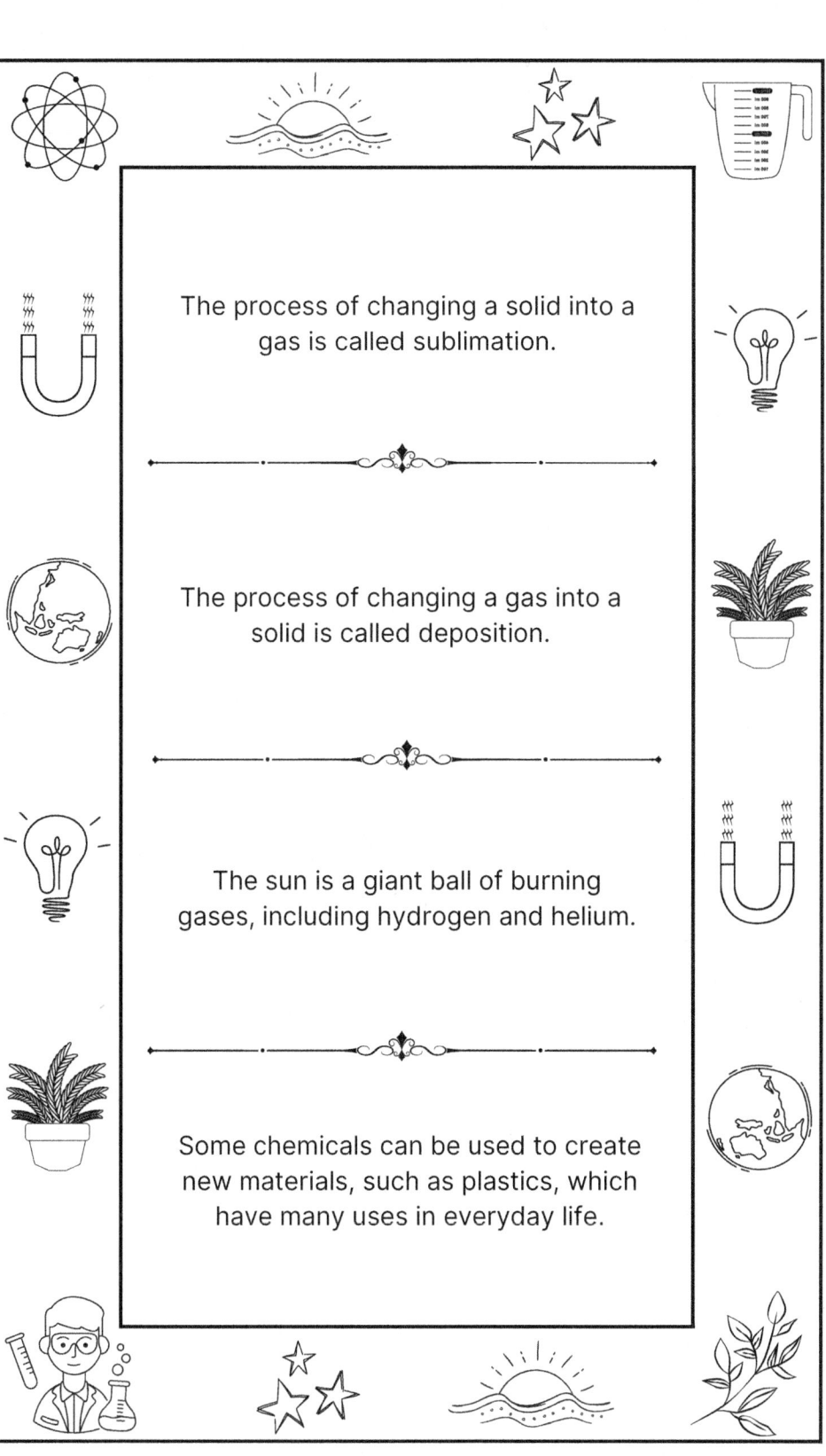

The process of changing a solid into a gas is called sublimation.

The process of changing a gas into a solid is called deposition.

The sun is a giant ball of burning gases, including hydrogen and helium.

Some chemicals can be used to create new materials, such as plastics, which have many uses in everyday life.

Some chemicals can change color when they react, which can be used to identify different substances.

The Earth's oceans are made up of saltwater.

Water freezes faster when it is warm and not cold.

Bee stings are acidic while wasp stings are alkaline.

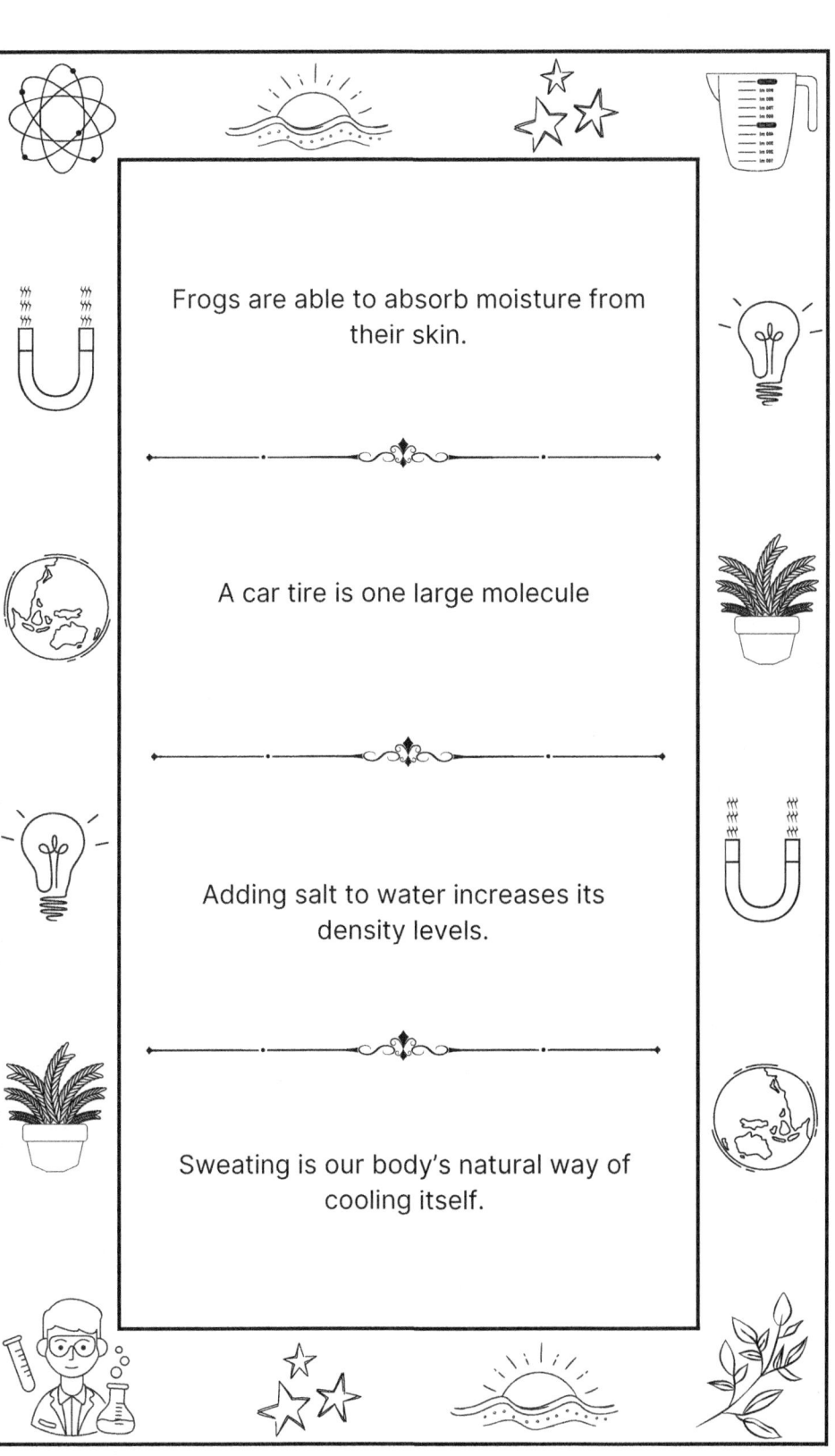

Frogs are able to absorb moisture from their skin.

A car tire is one large molecule

Adding salt to water increases its density levels.

Sweating is our body's natural way of cooling itself.

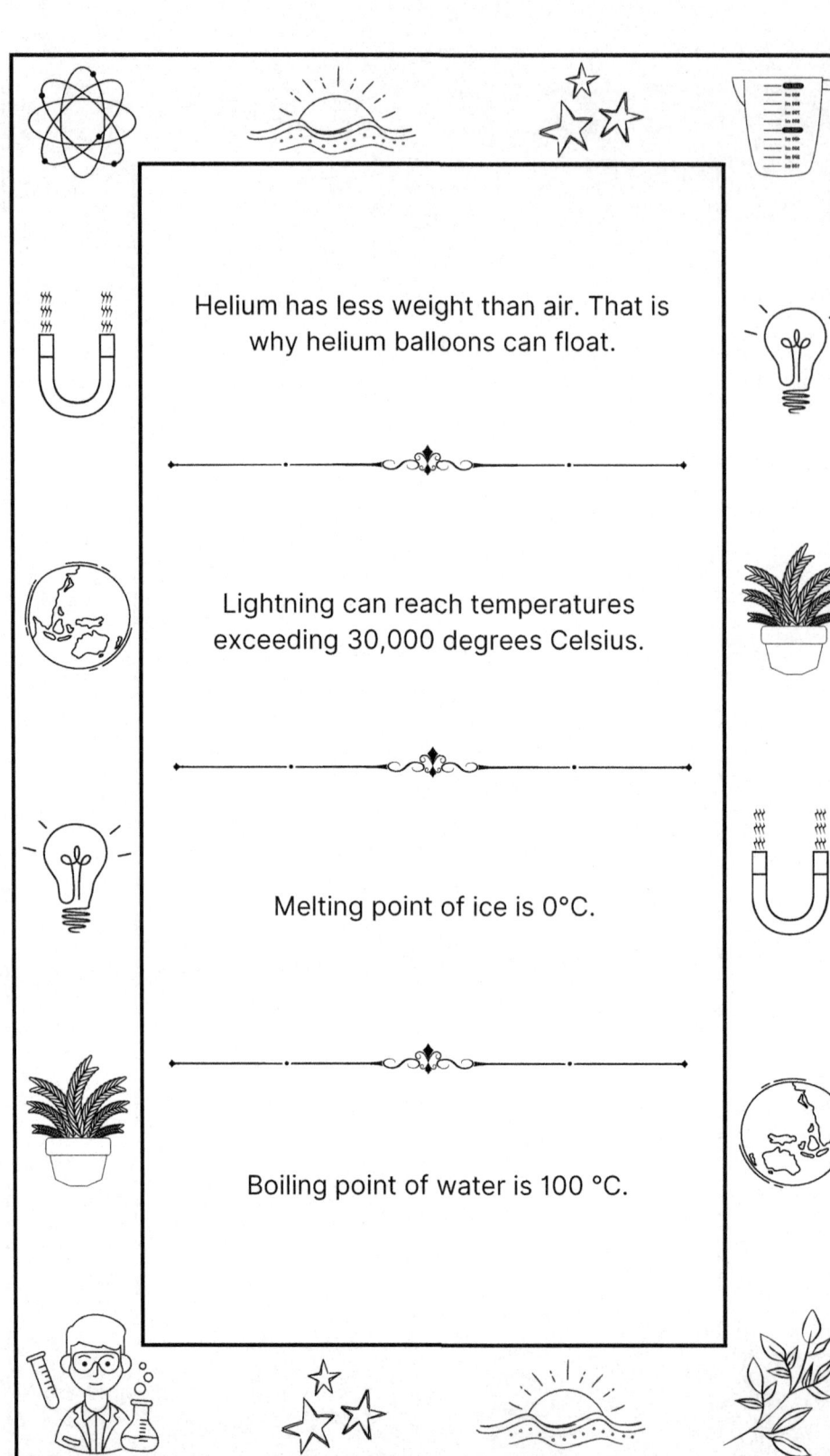

Helium has less weight than air. That is why helium balloons can float.

Lightning can reach temperatures exceeding 30,000 degrees Celsius.

Melting point of ice is 0°C.

Boiling point of water is 100 °C.

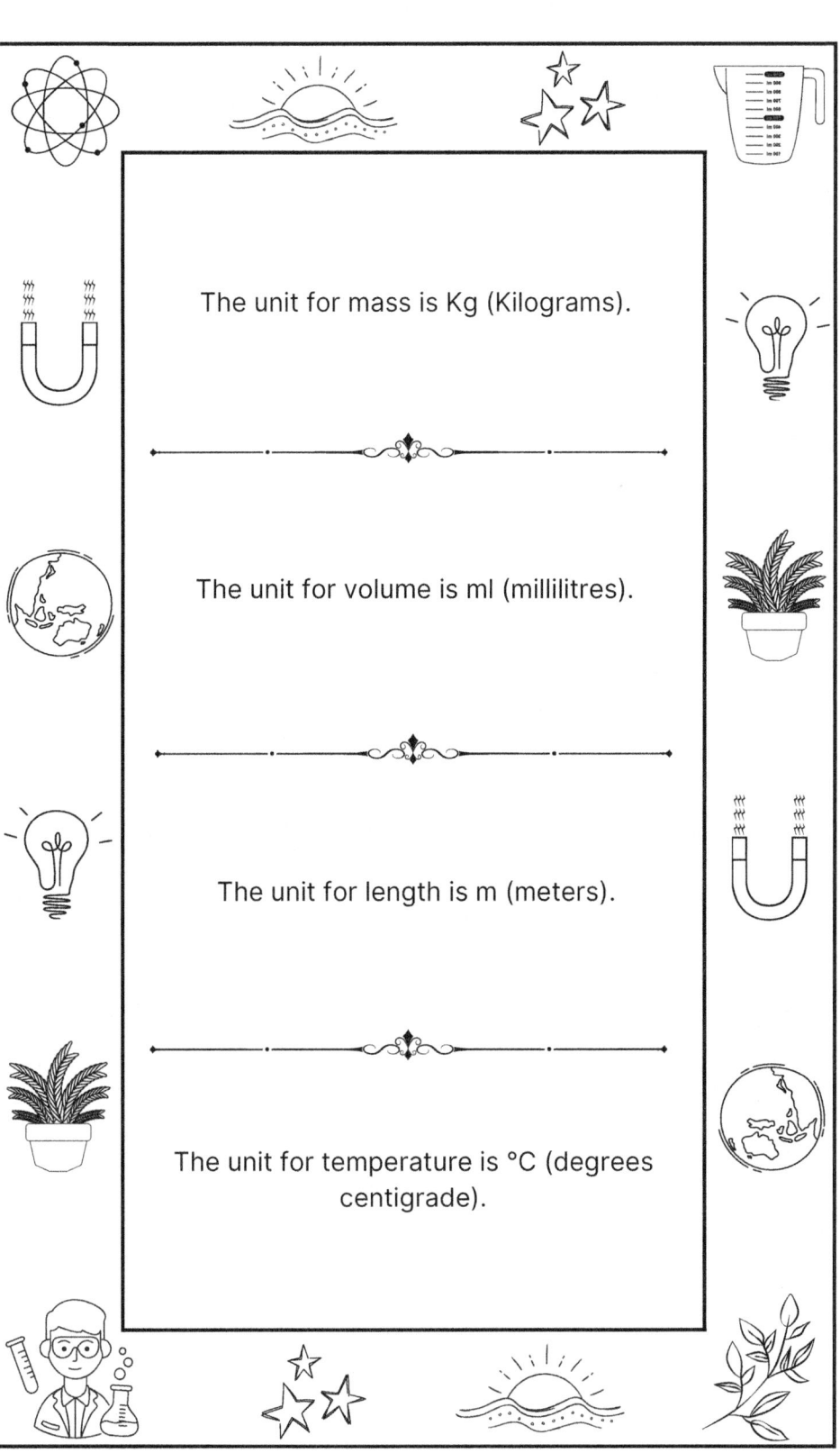

The unit for mass is Kg (Kilograms).

The unit for volume is ml (millilitres).

The unit for length is m (meters).

The unit for temperature is °C (degrees centigrade).

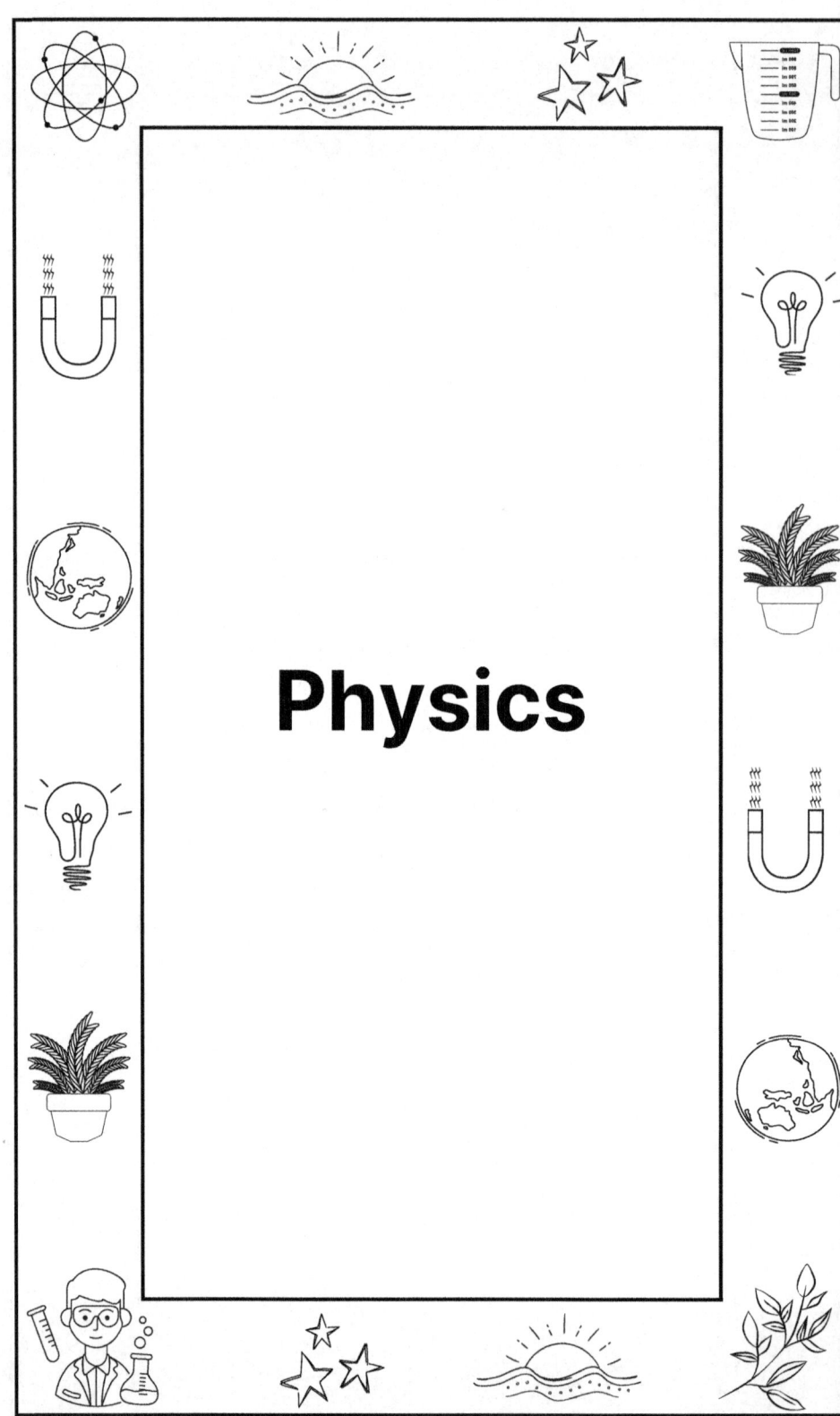

Physics

Physics is the study of the basic properties and laws of matter and energy.

The goal of physics is to understand and predict how the world works.

Isaac Newton came up with three laws that explain how things move.

The first law says that things will stay still unless something pushes or pulls on them.

The second law says that the harder you push or pull on something, the faster it will go.

The third law says that for every action, there is an equal and opposite reaction.

Matter and energy are the two things that the entire universe is made of.

Light travels in a straight line unless something gets in its way.

Sound travels at a speed of around 767 miles per hour.

Sound is caused by vibrations in the air.

The sky is blue because the atmosphere scatters blue light from the sun.

The speed of light is a very important number in physics, and it is always the same no matter how fast you are moving.

A scientist who studies physics is known as a physicist.

Light from the Earth takes just 1.255 seconds to reach the Moon.

Because of differences in gravity, a 200 pound person would only weigh 76 pounds on Mars.

Electric eels uses its shock to stun prey and keep predators at bay.

There are four fundamental forces in the universe: gravity, electromagnetism, the strong nuclear force, and the weak nuclear force.

Gravity is the force that makes things fall towards the ground.

Electromagnetism is the force that makes some things push away or pull towards each other.

The strong nuclear force is the force that holds the tiny particles inside an atom's nucleus together.

The weak nuclear force is the force that makes some types of atoms decay, or fall apart.

Thermodynamics is the study of heat, energy, and work.

Light is the only form of energy that we can actually see with our naked eyes.

It takes sunlight an average of 8 minutes and 20 seconds to travel from the Sun to the Earth.

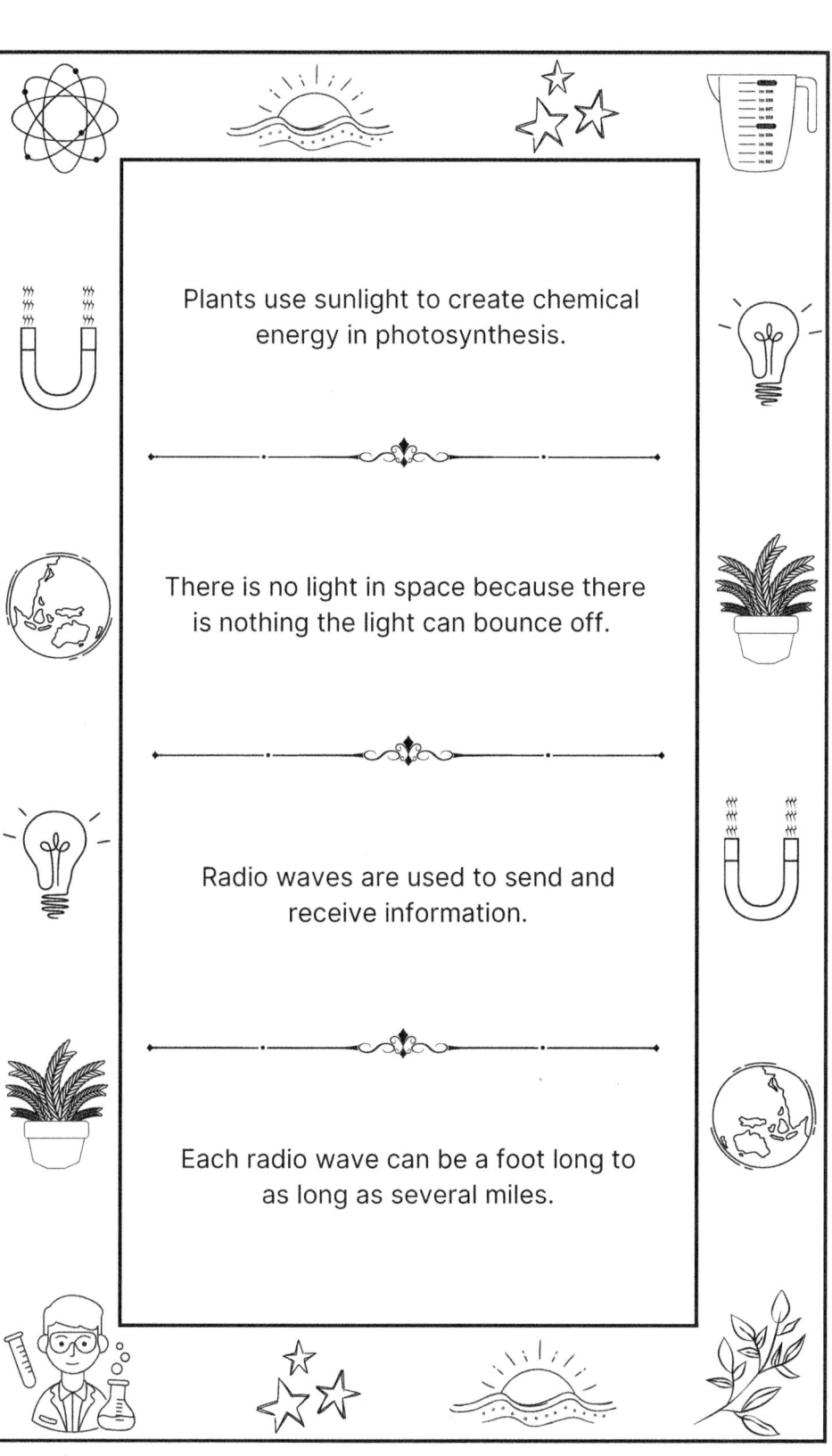

Plants use sunlight to create chemical energy in photosynthesis.

There is no light in space because there is nothing the light can bounce off.

Radio waves are used to send and receive information.

Each radio wave can be a foot long to as long as several miles.

Gamma rays have the greatest energy.

There is no gravity in outer space.

The Lockheed SR-71 Blackbird is the fastest jet aircraft in the world.
It can travel at a speed of about 3,500 kph (2,100 mph).

The oxygen in the airplane emergency masks lasts only for about 15 minutes.

In 1947, Chuck Yeager became the first person to fly at a supersonic speed.

The average speed of a commercial aircraft is about 800 kilometers per hour.

Radiation is the transfer of heat as waves.

Electromagnets are used by some high-speed trains.

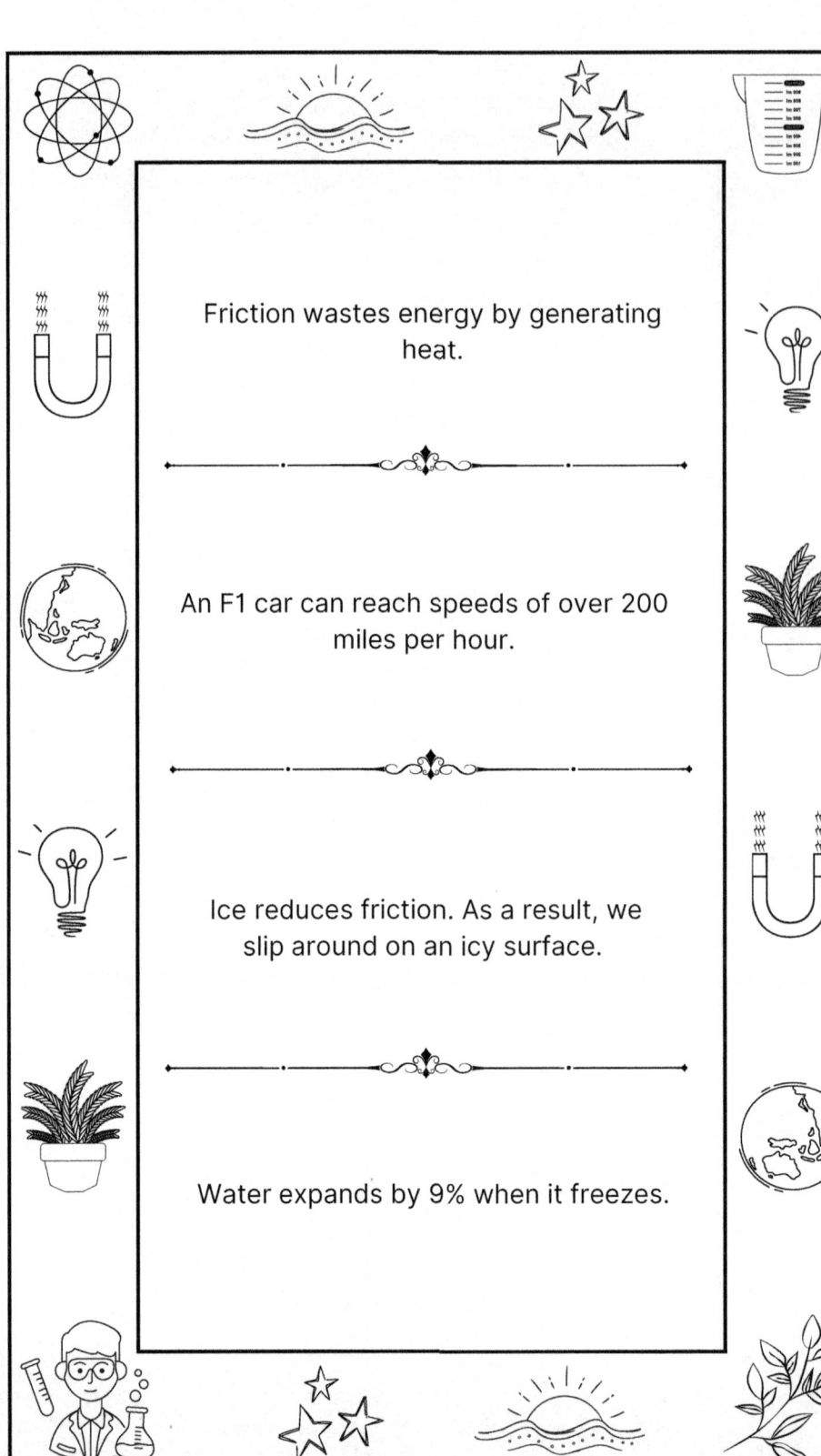

Friction wastes energy by generating heat.

An F1 car can reach speeds of over 200 miles per hour.

Ice reduces friction. As a result, we slip around on an icy surface.

Water expands by 9% when it freezes.

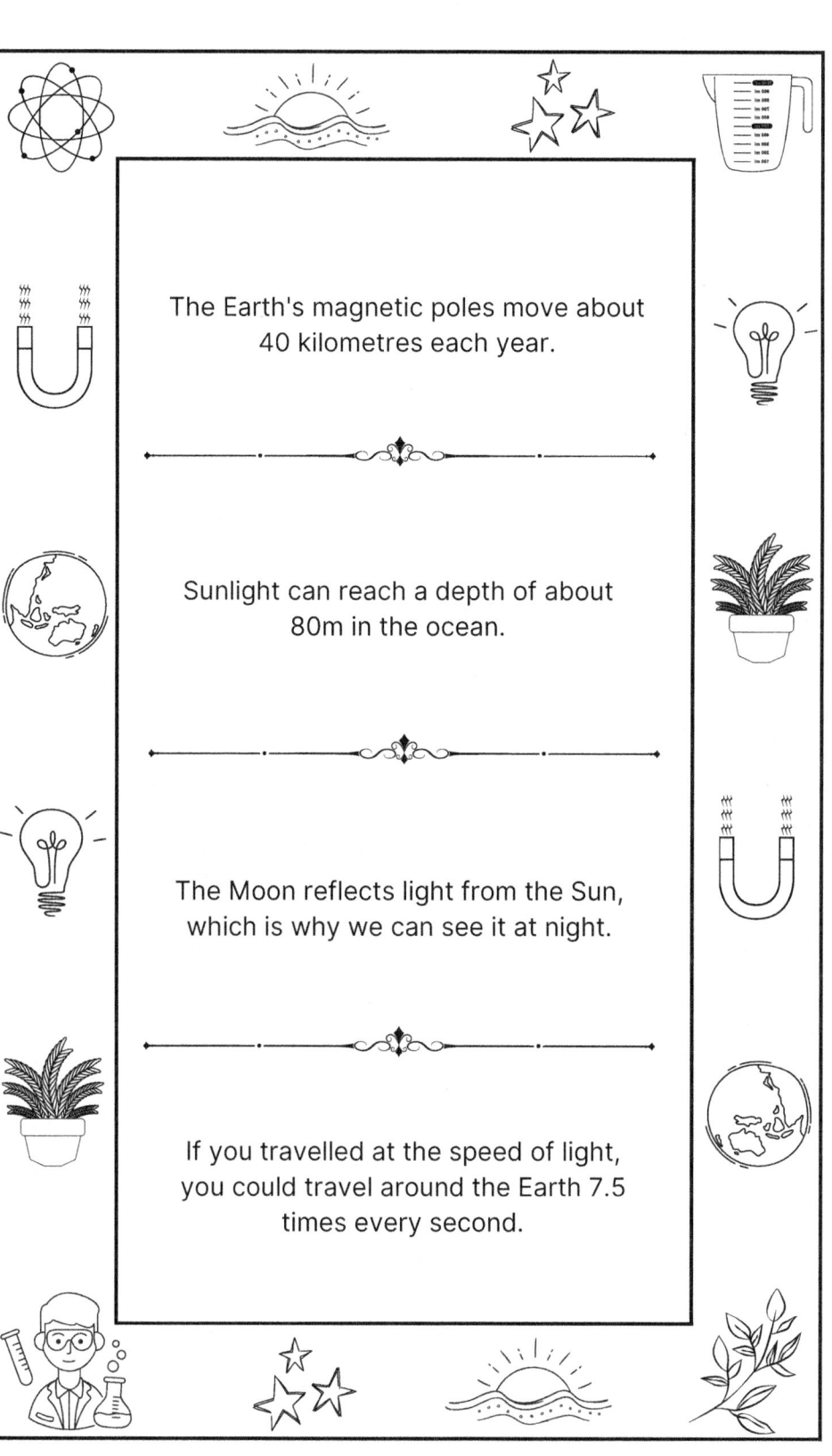

The Earth's magnetic poles move about 40 kilometres each year.

Sunlight can reach a depth of about 80m in the ocean.

The Moon reflects light from the Sun, which is why we can see it at night.

If you travelled at the speed of light, you could travel around the Earth 7.5 times every second.

Animals that only come out at night when it is dark are called nocturnal animals.

Hot water freezes faster than cold water.

Physics is involved in many aspects of our daily lives, such as the design and operation of vehicles, communication technology and medical equipment.

Ocean tides are caused by the gravity of the moon.

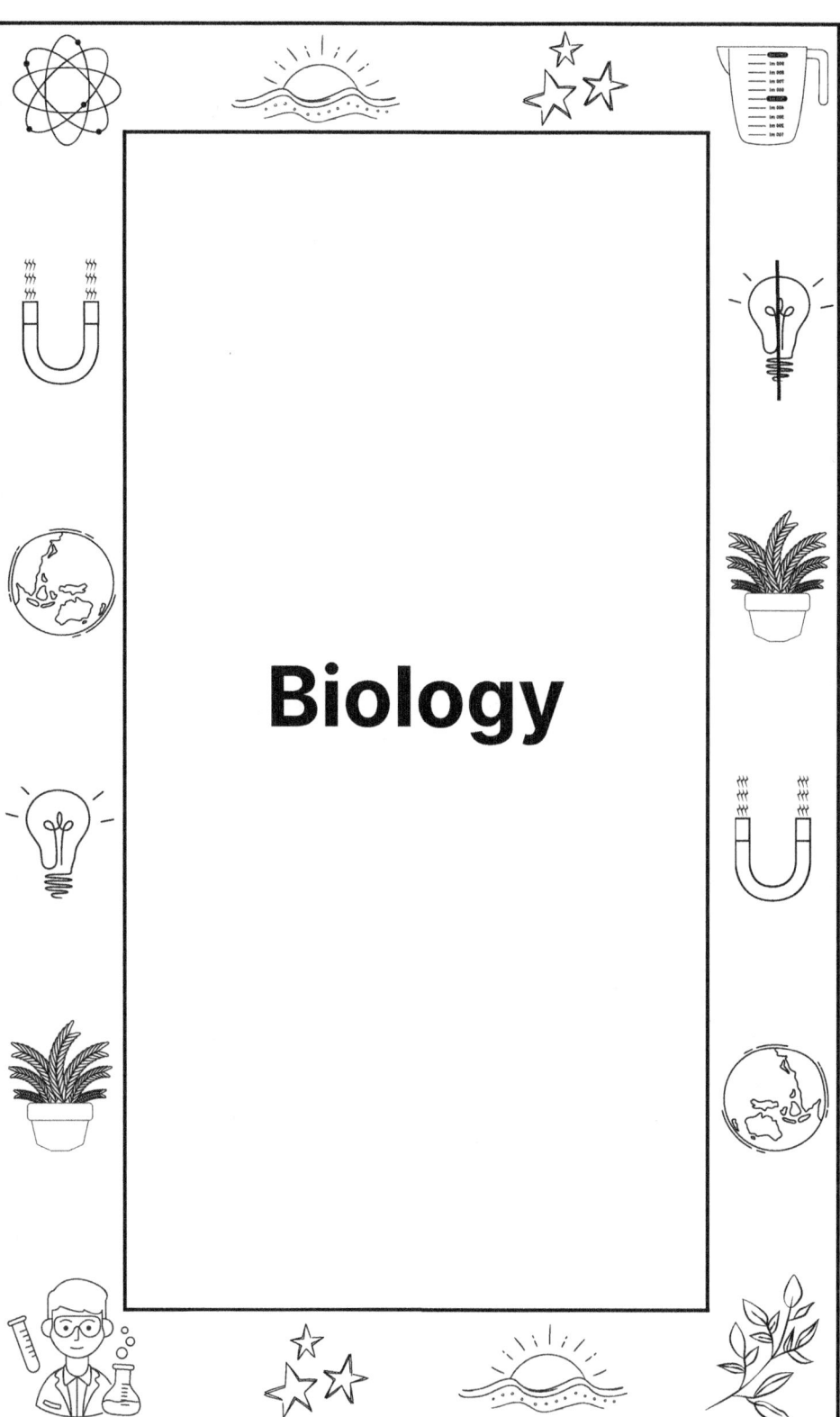

Biology

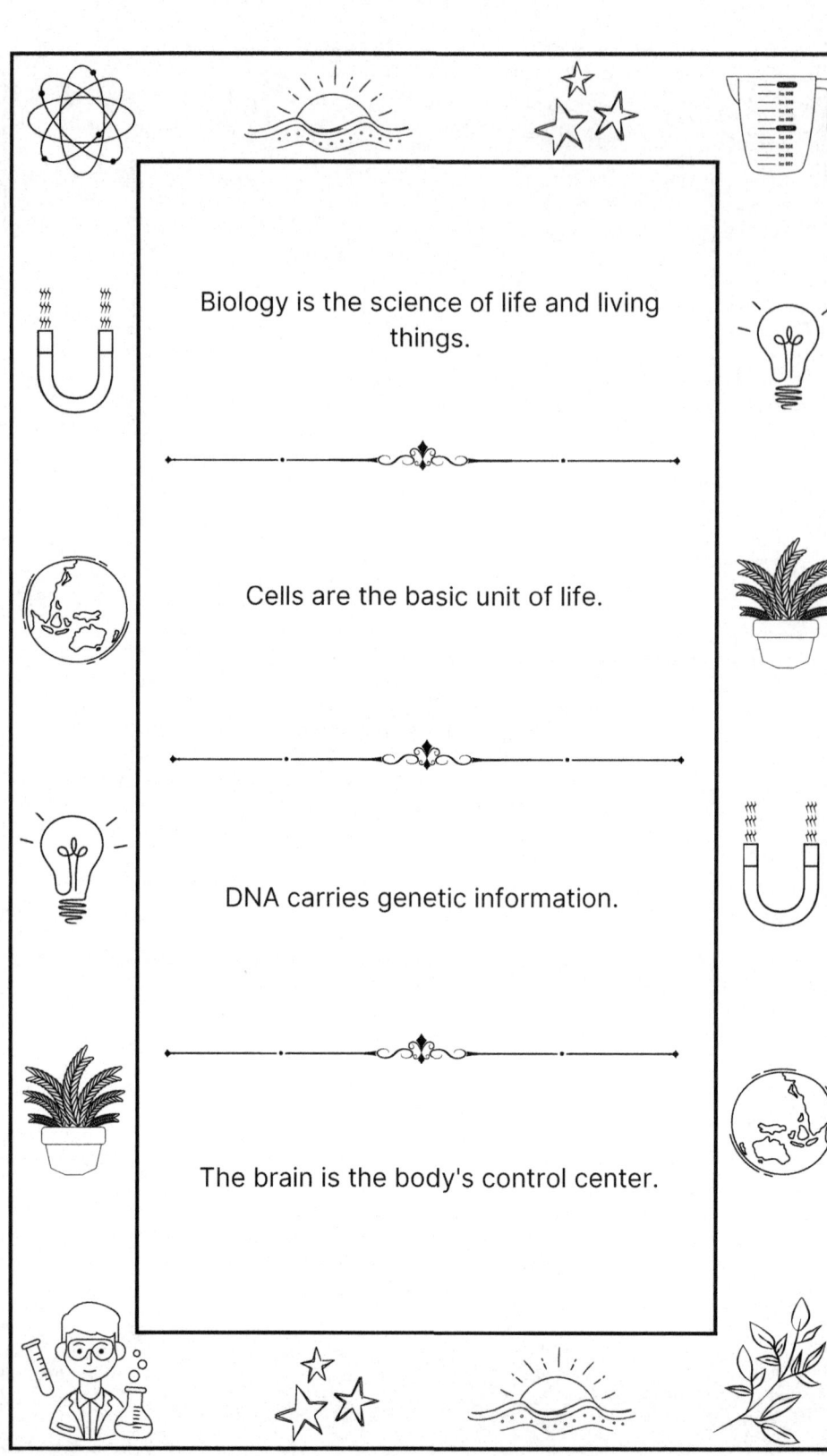

Biology is the science of life and living things.

Cells are the basic unit of life.

DNA carries genetic information.

The brain is the body's control center.

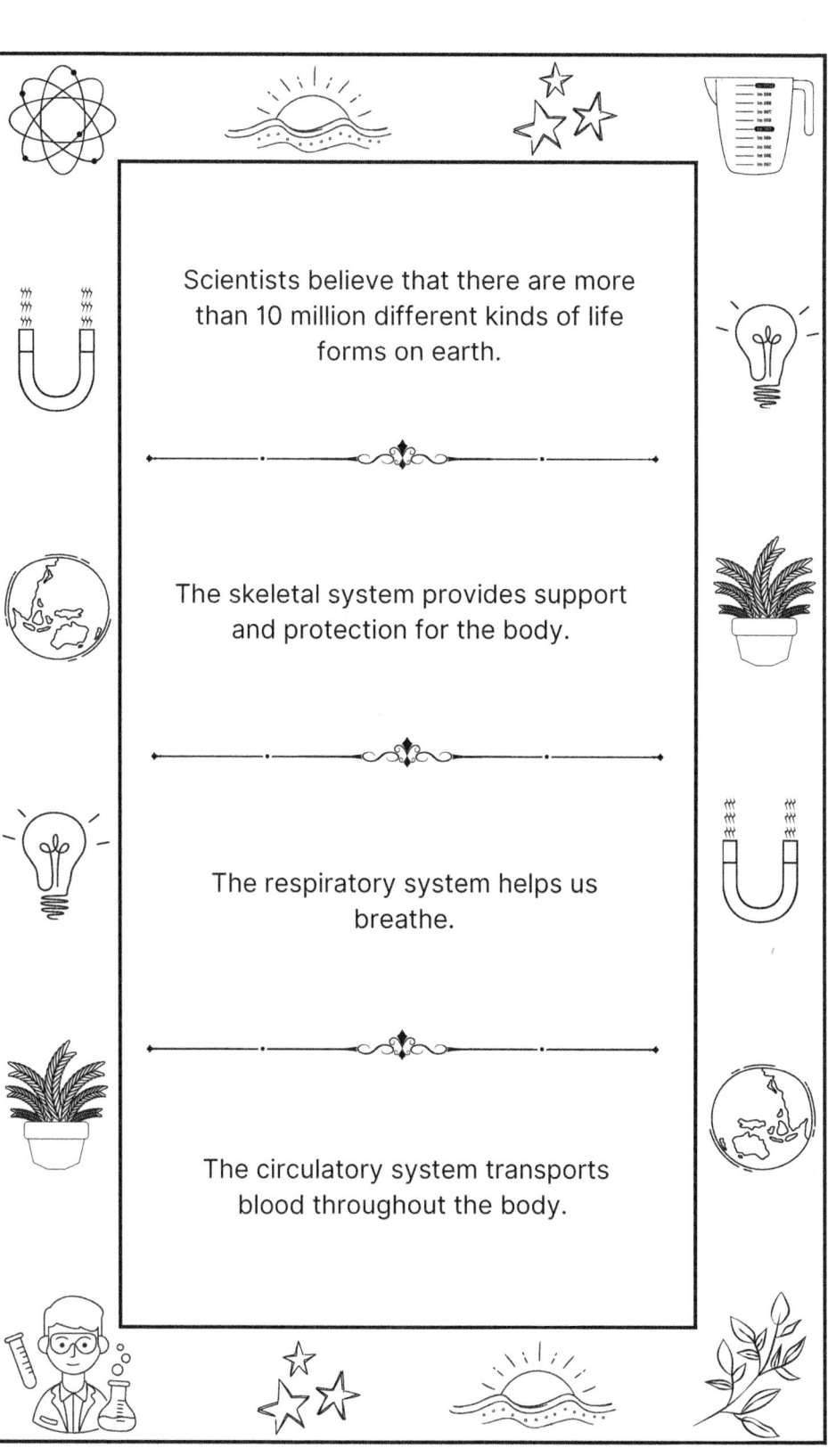

Scientists believe that there are more than 10 million different kinds of life forms on earth.

The skeletal system provides support and protection for the body.

The respiratory system helps us breathe.

The circulatory system transports blood throughout the body.

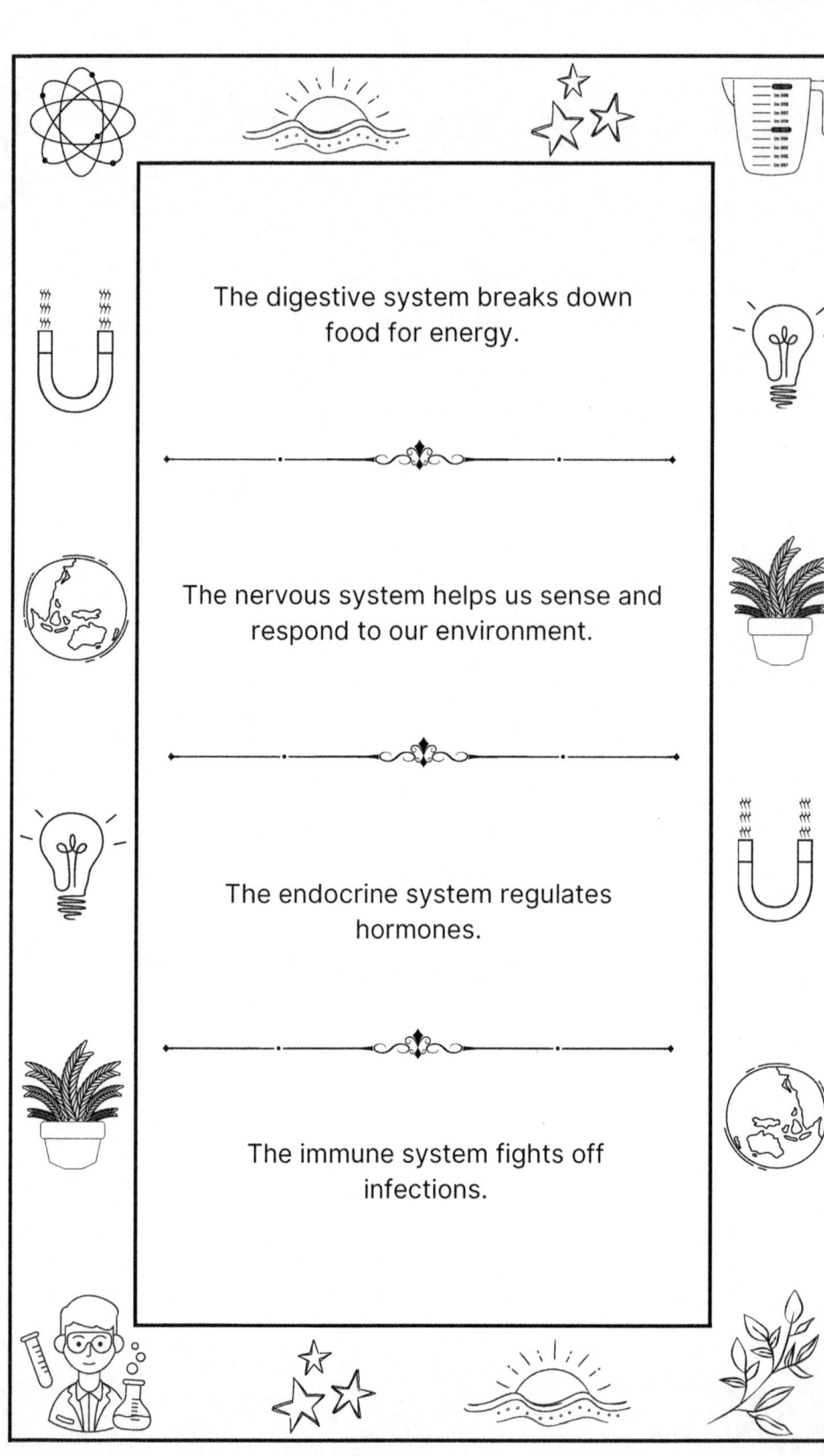

The digestive system breaks down food for energy.

The nervous system helps us sense and respond to our environment.

The endocrine system regulates hormones.

The immune system fights off infections.

The urinary system removes waste from the body.

The muscular system helps us move.

The skin protects the body and helps regulate body temperature.

The five senses are sight, hearing, taste, smell, and touch.

Mammals have hair or fur.

Reptiles are cold-blooded animals.

Birds have feathers and beaks.

Fish live in water and breathe through gills.

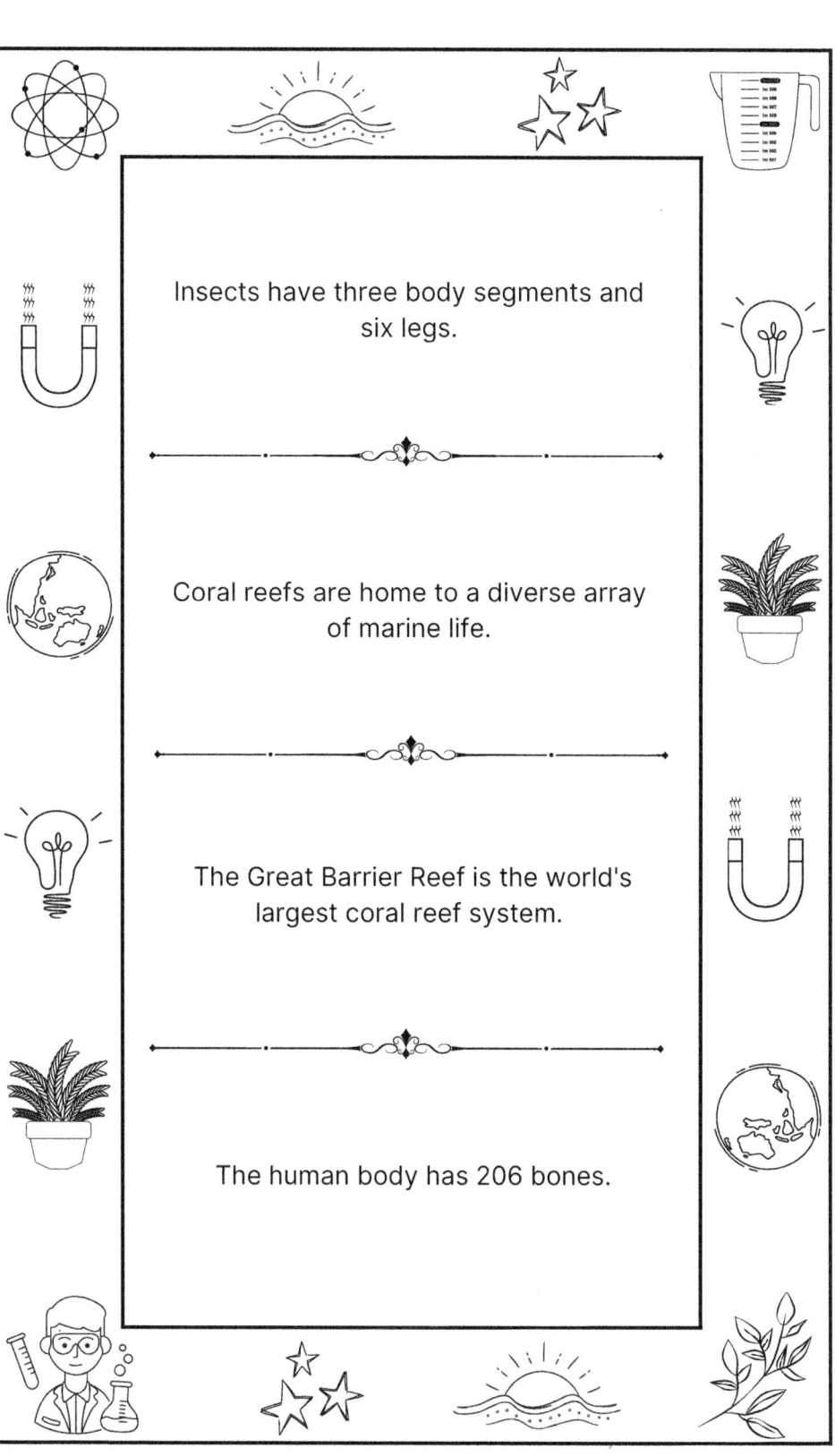

Insects have three body segments and six legs.

Coral reefs are home to a diverse array of marine life.

The Great Barrier Reef is the world's largest coral reef system.

The human body has 206 bones.

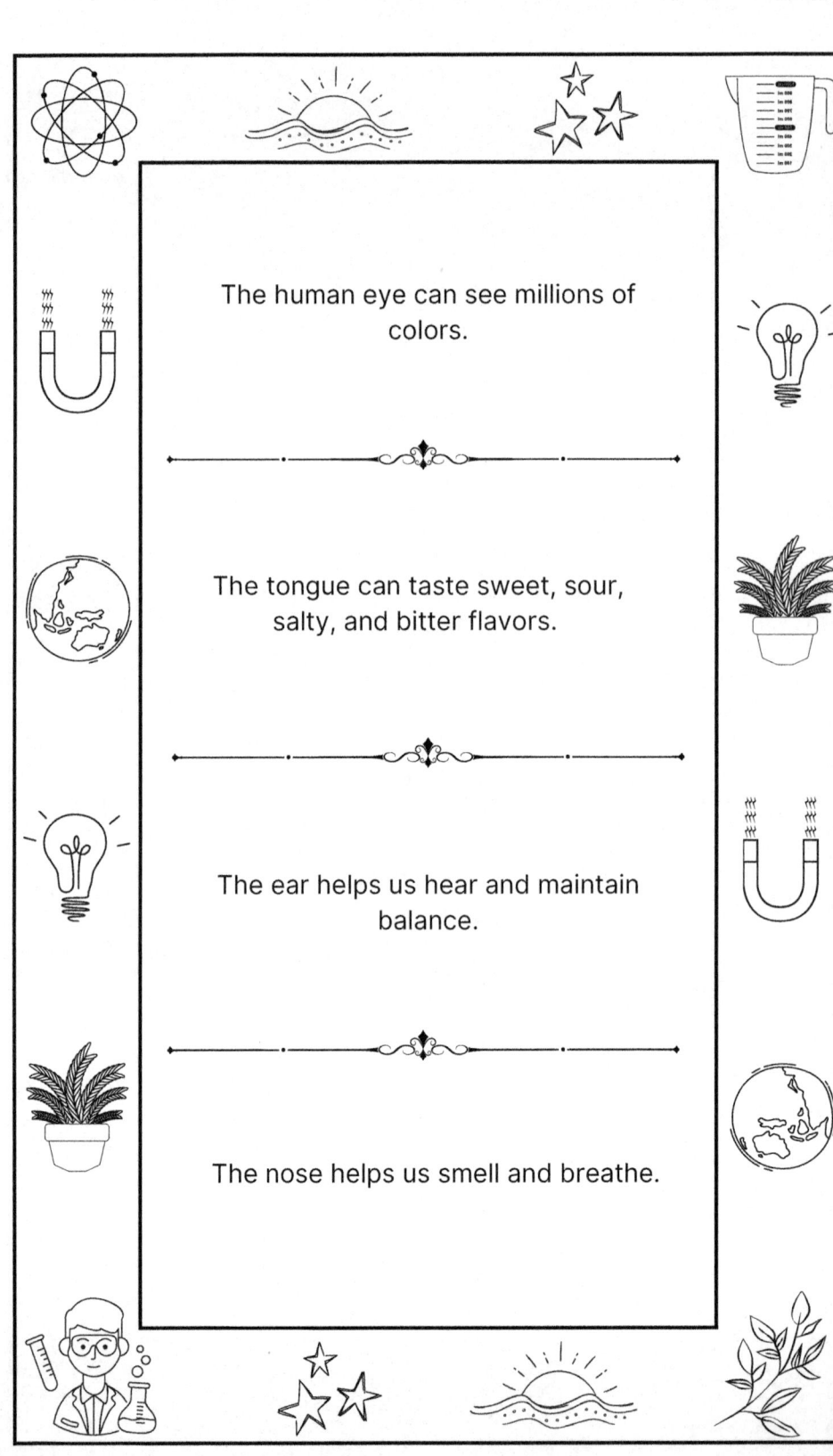

The human eye can see millions of colors.

The tongue can taste sweet, sour, salty, and bitter flavors.

The ear helps us hear and maintain balance.

The nose helps us smell and breathe.

The pancreas produces insulin to regulate blood sugar.

The liver filters toxins from the blood.

The kidneys filter waste from the blood.

The thyroid gland regulates metabolism.

The adrenal gland produces hormones like adrenaline.

The pituitary gland produces hormones that regulate growth and metabolism.

People with higher IQ dream more.

An average human being may not remember 90% of their dreams.

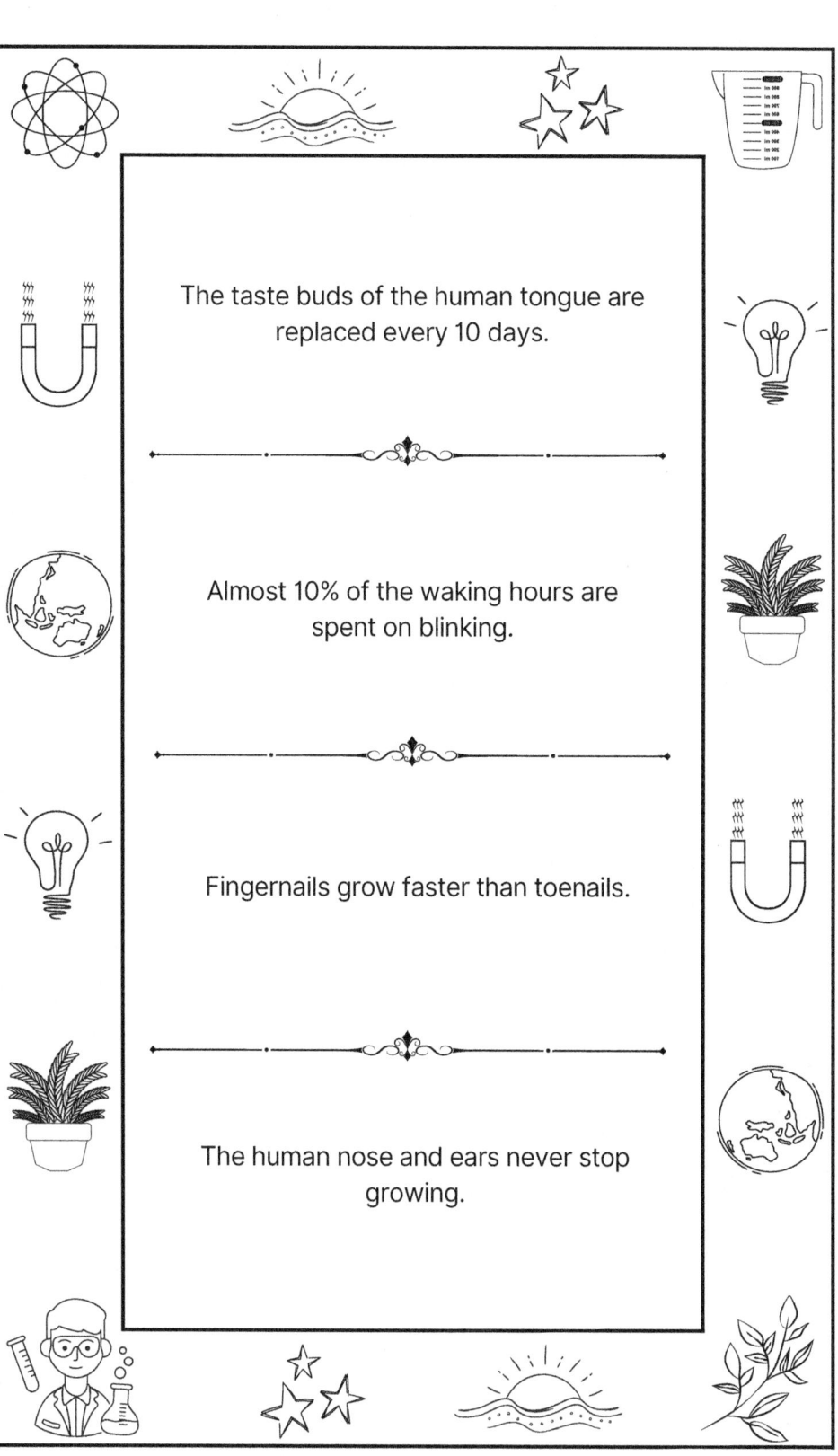

The taste buds of the human tongue are replaced every 10 days.

Almost 10% of the waking hours are spent on blinking.

Fingernails grow faster than toenails.

The human nose and ears never stop growing.

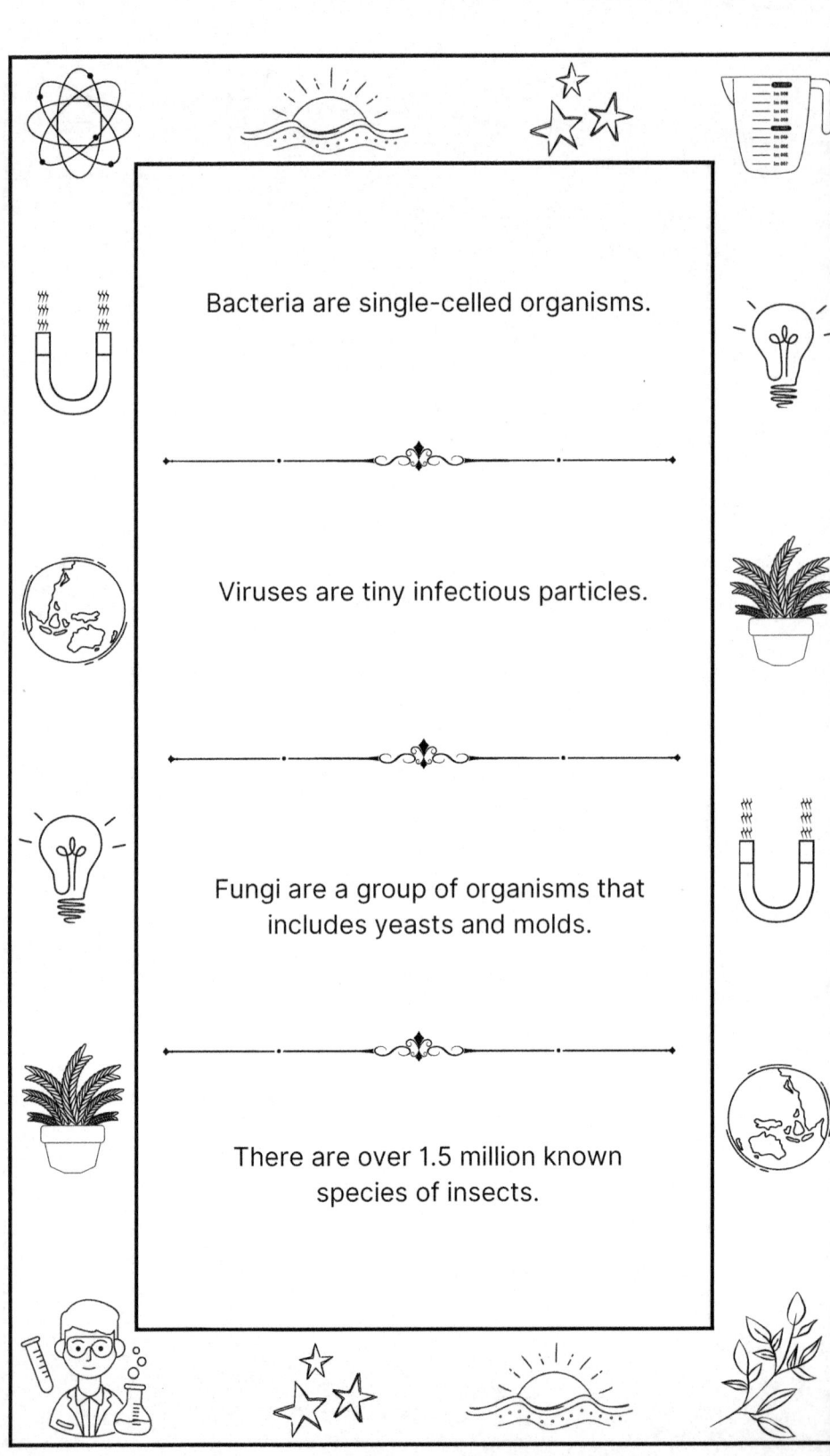

Bacteria are single-celled organisms.

Viruses are tiny infectious particles.

Fungi are a group of organisms that includes yeasts and molds.

There are over 1.5 million known species of insects.

The smallest mammal in the world is the bumblebee bat, which weighs less than a penny.

The tallest mammal in the world is the giraffe, which can grow up to 18 feet tall.

Humans and gorillas share about 98% of the same DNA.

The human nose can detect over 1 trillion different scents.

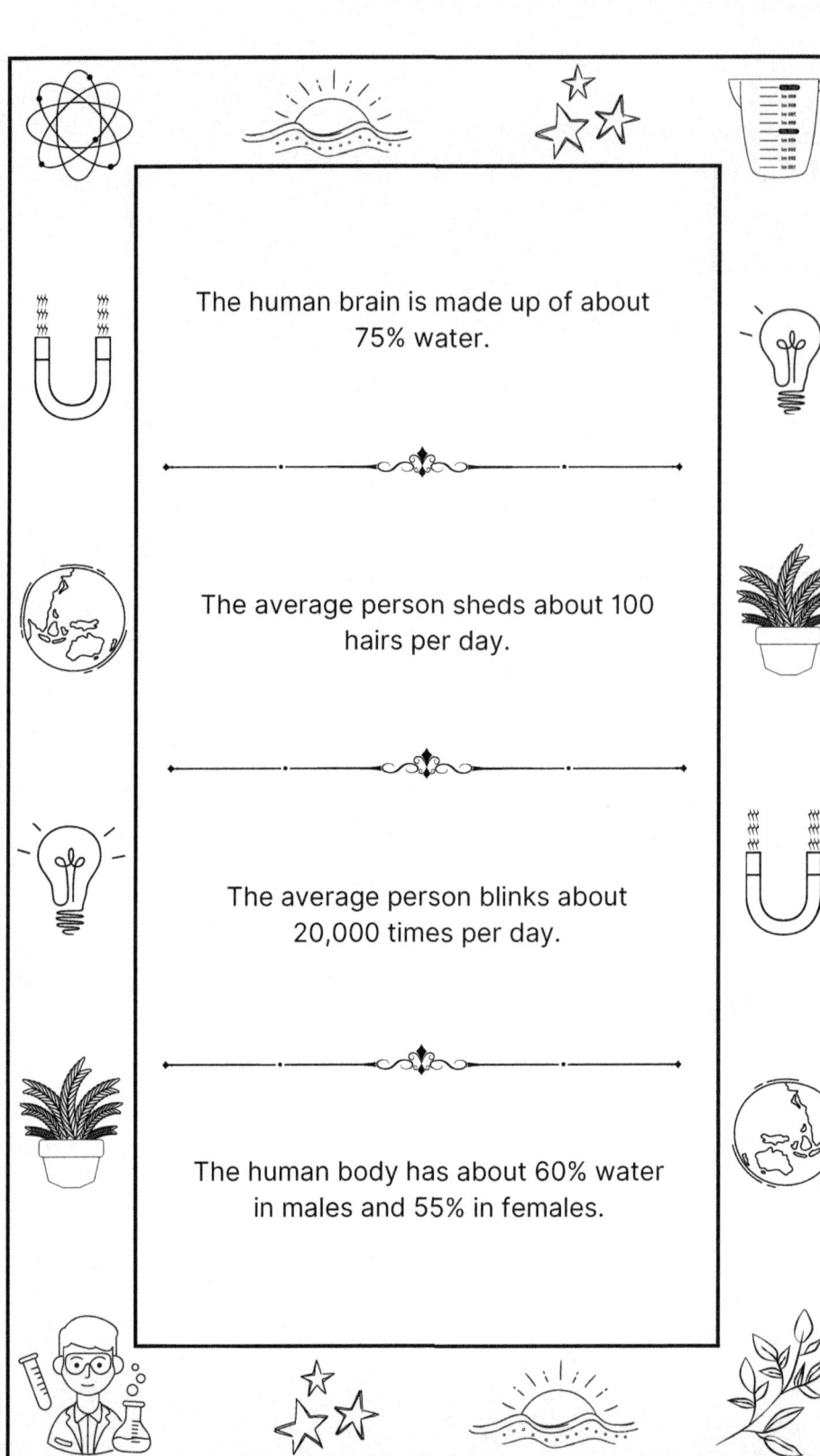

The human brain is made up of about 75% water.

The average person sheds about 100 hairs per day.

The average person blinks about 20,000 times per day.

The human body has about 60% water in males and 55% in females.

The average person has about 100,000 hairs on their head.

The human body is made up of around 37 trillion cells.

The smallest bone in the human body is the stapes, which is located in the ear.

The longest bone in the human body is the femur, which is located in the thigh.

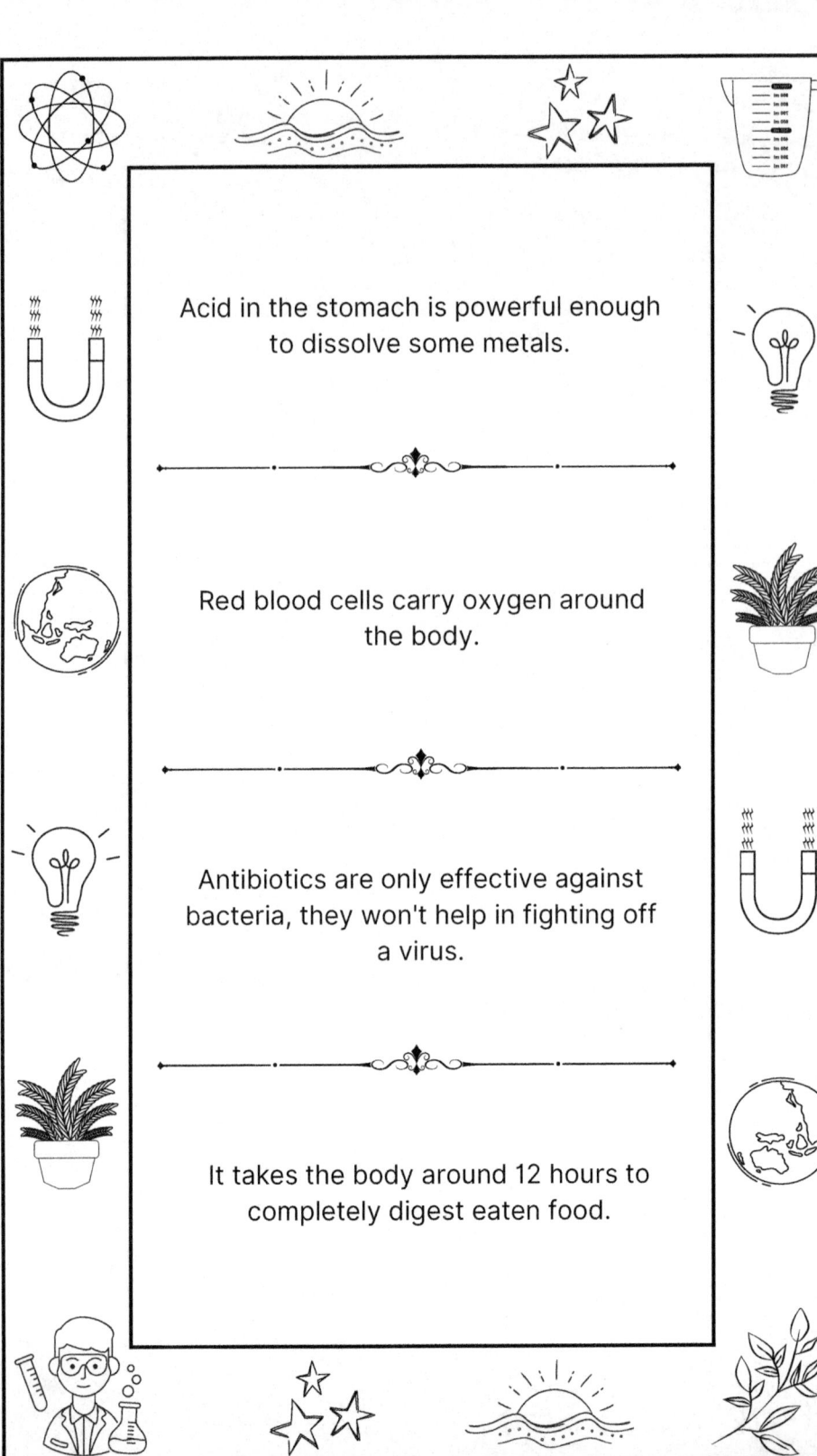

Acid in the stomach is powerful enough to dissolve some metals.

Red blood cells carry oxygen around the body.

Antibiotics are only effective against bacteria, they won't help in fighting off a virus.

It takes the body around 12 hours to completely digest eaten food.

As well as having unique fingerprints, humans also have unique tongue prints.

Your nose and ears continue growing throughout your entire life.

Most adults have 32 teeth.

Men blink half as often as women.

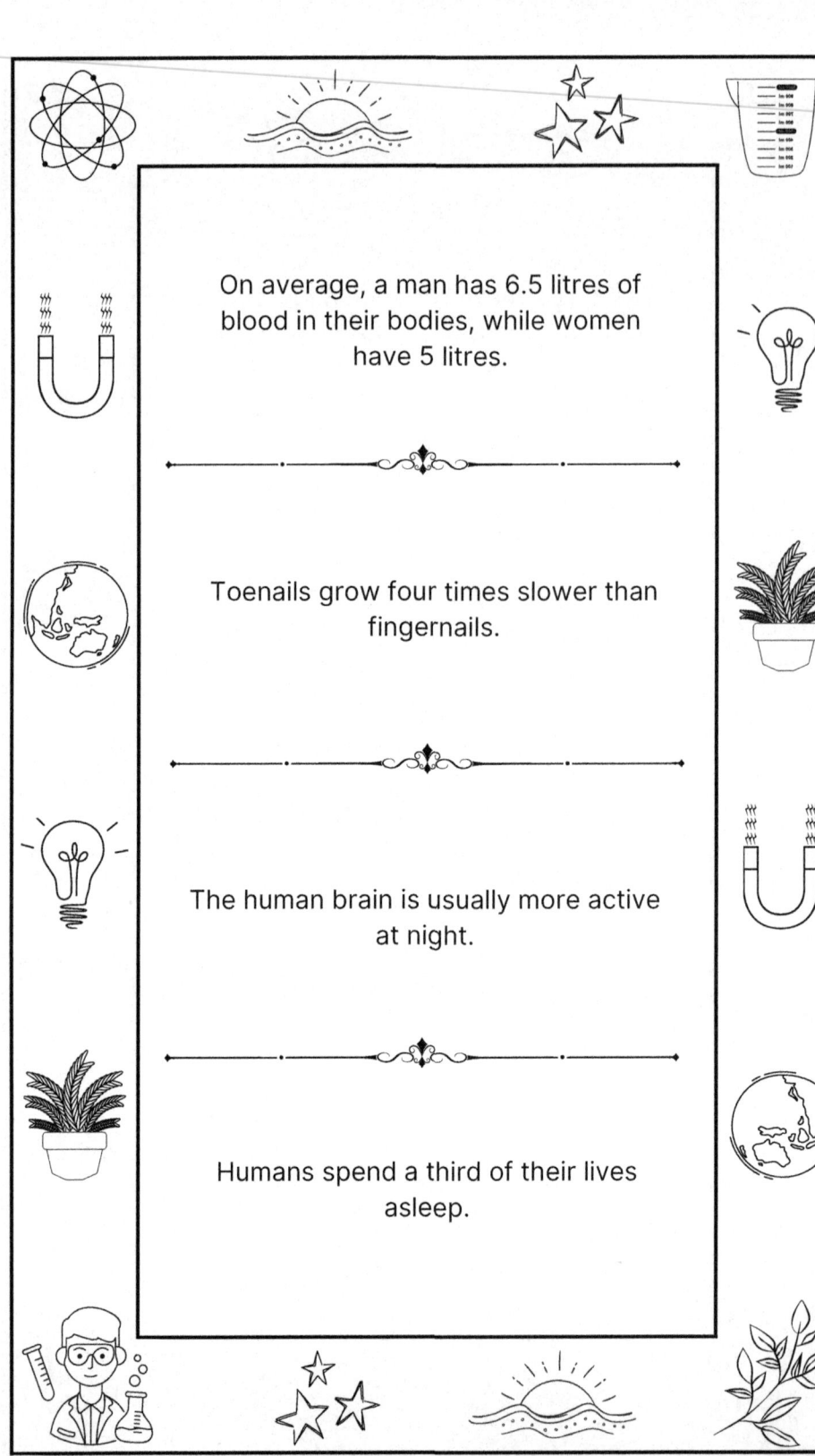

On average, a man has 6.5 litres of blood in their bodies, while women have 5 litres.

Toenails grow four times slower than fingernails.

The human brain is usually more active at night.

Humans spend a third of their lives asleep.

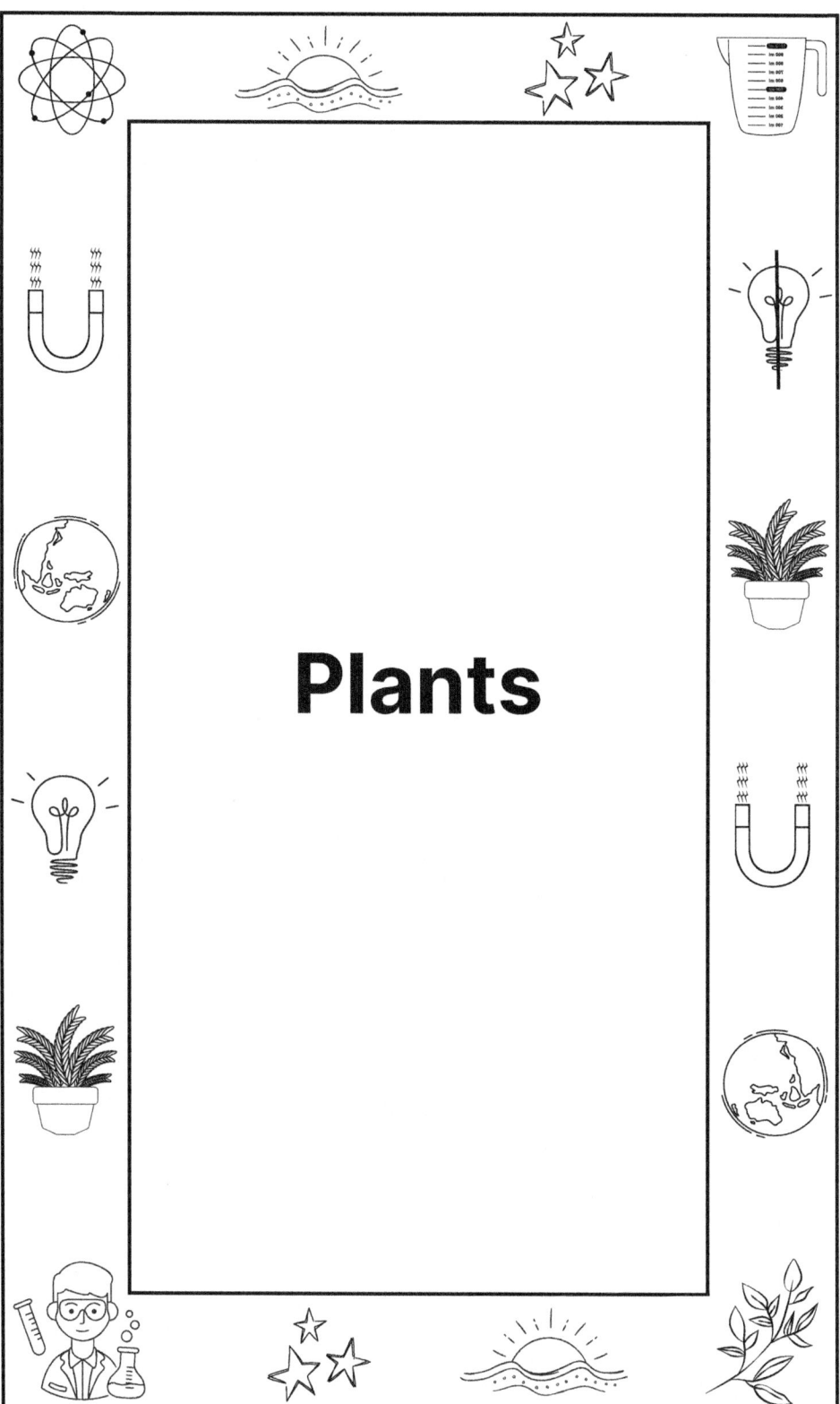

Plants

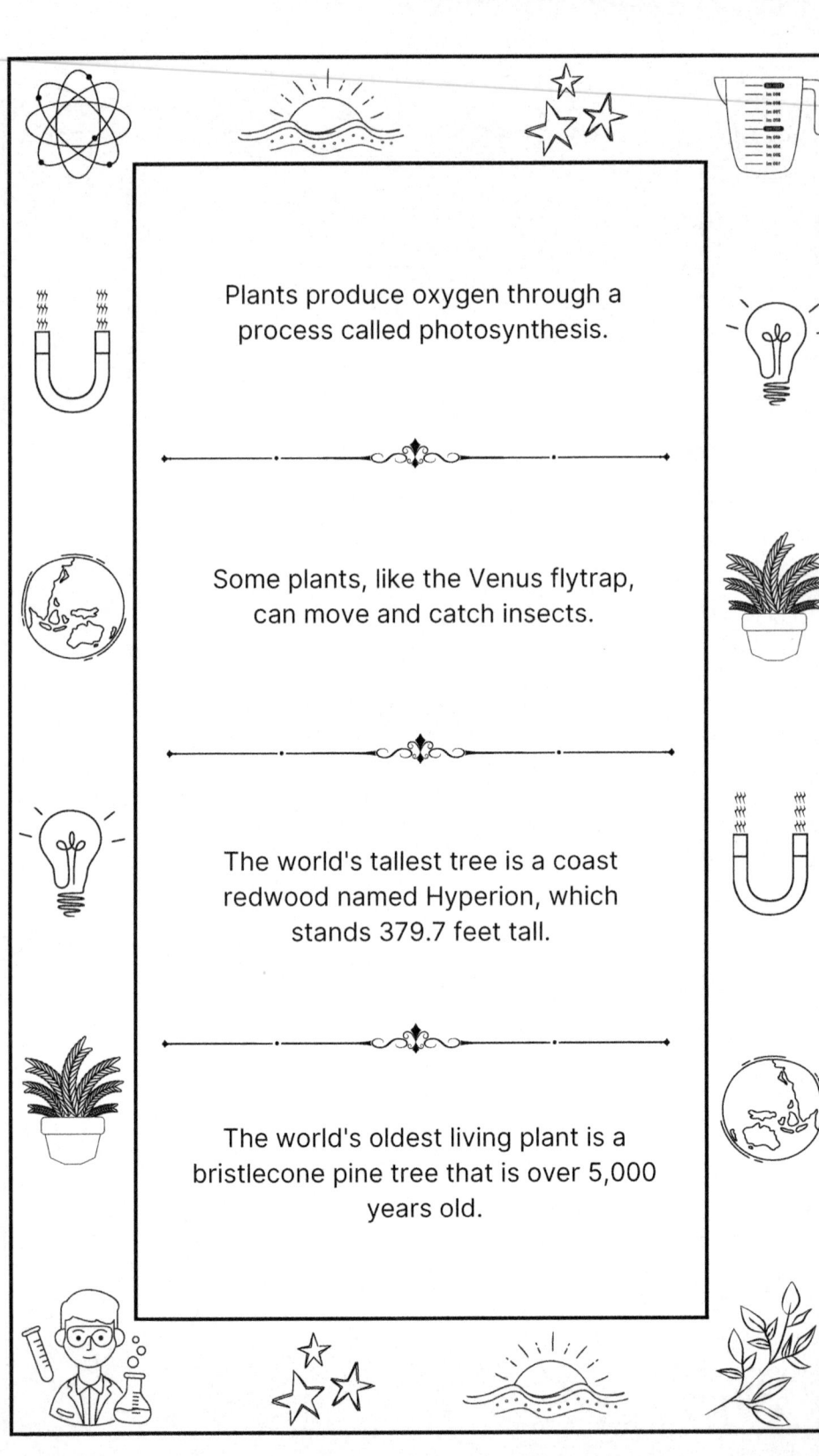

Plants produce oxygen through a process called photosynthesis.

Some plants, like the Venus flytrap, can move and catch insects.

The world's tallest tree is a coast redwood named Hyperion, which stands 379.7 feet tall.

The world's oldest living plant is a bristlecone pine tree that is over 5,000 years old.

Bamboo is the fastest growing plant in the world, with some species growing up to 3 feet in a single day.

Some plants, such as the cactus, have adapted to survive in dry, desert environments

---

The Amazon rainforest is home to over 40,000 plant species.

The world's largest flower is the Rafflesia arnoldii, which can grow up to 3 feet wide and 10 pounds.

Many plants, such as tomatoes and peppers, are actually fruit.

Lotus can grow in very dirty water and still be safe to eat.

Many plants produce scents to attract pollinators or deter herbivores.

The Moringa tree is known as the "miracle tree" because almost every part of it is edible or has medicinal properties.

The baobab tree can store up to 120,000 liters of water in its trunk.

The cacao plant is the source of cocoa beans, which are used to make chocolate.

The coffee plant is the source of coffee beans.

Some plants, such as the nepenthes, can trap and digest insects for nutrients.

The pitcher plant is a carnivorous plant that traps insects in a sticky liquid inside its pitcher-shaped leaves.

The saguaro cactus is the largest cactus in the United States, and can grow up to 50 feet tall.

The banyan tree is a type of fig tree that can grow roots from its branches to support itself and form a "forest" of trees.

80% of flowering plants are adapted for pollination by animals (mostly insects).

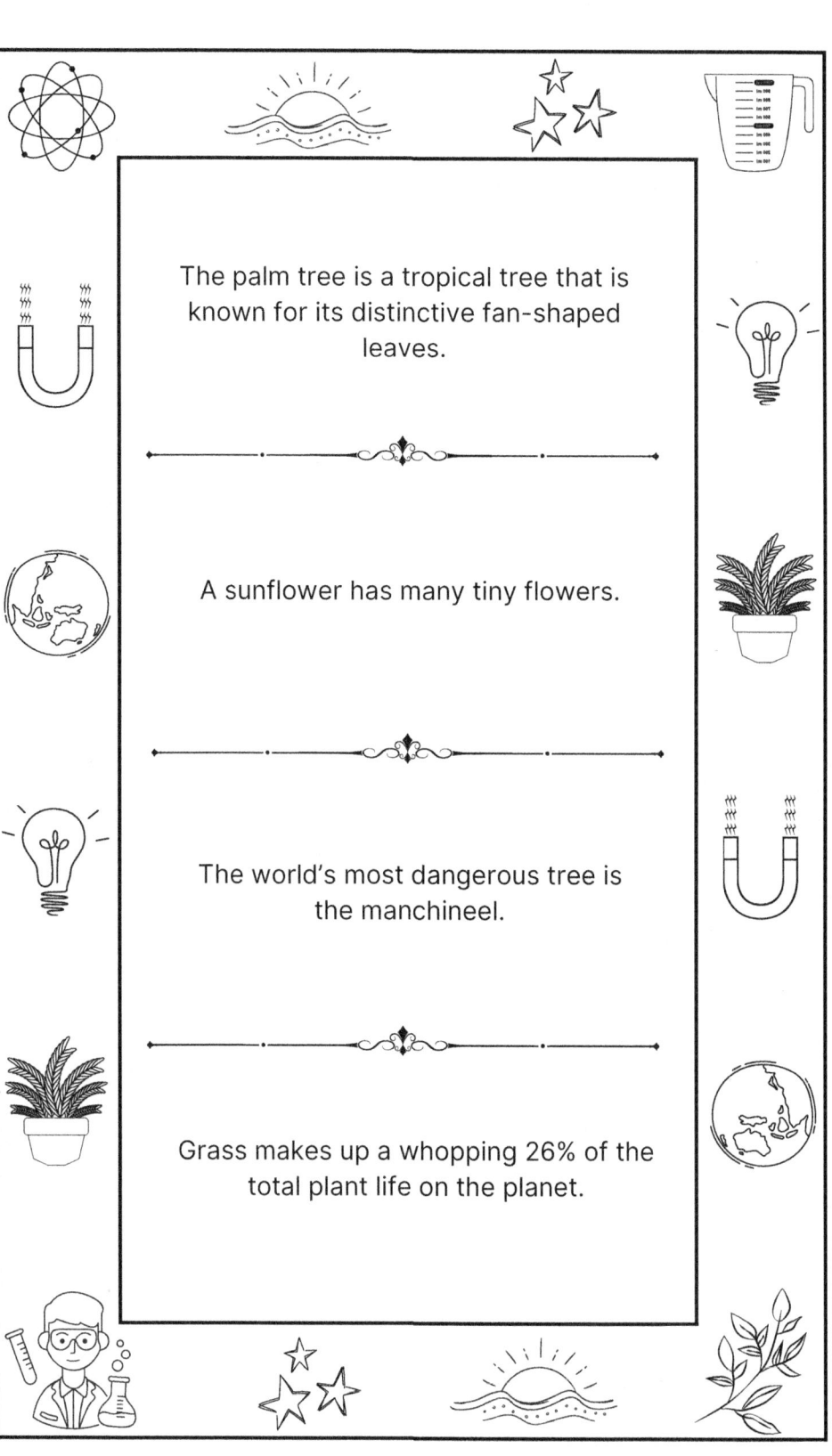

The palm tree is a tropical tree that is known for its distinctive fan-shaped leaves.

A sunflower has many tiny flowers.

The world's most dangerous tree is the manchineel.

Grass makes up a whopping 26% of the total plant life on the planet.

Fir trees are natural Christmas trees.

Brazil is the only country named after a tree.

The food humans eat comes from just about 30 plants.

Leaves are used for taking light from the Sun which is used in the process of photosynthesis.

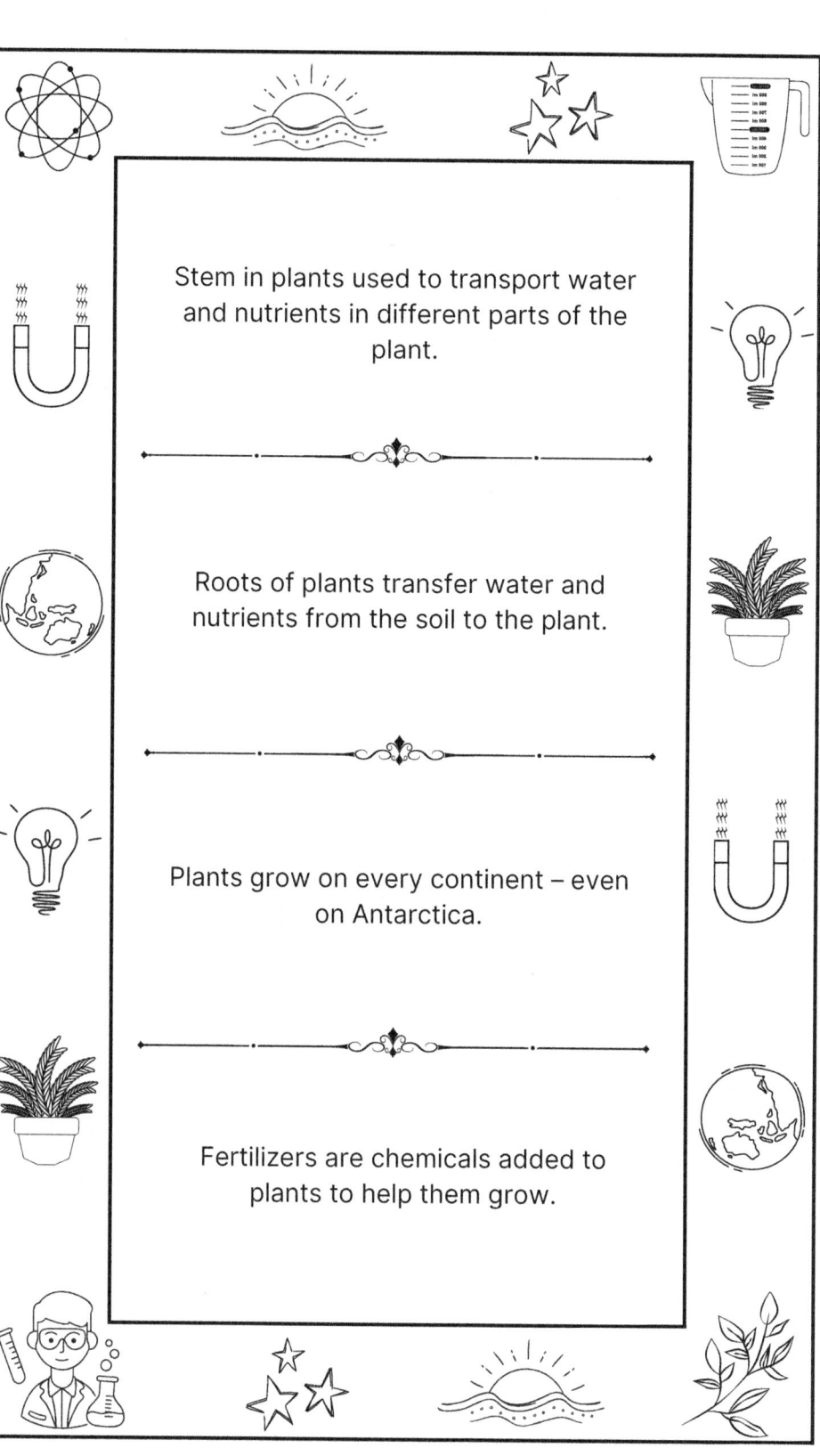

Stem in plants used to transport water and nutrients in different parts of the plant.

Roots of plants transfer water and nutrients from the soil to the plant.

Plants grow on every continent – even on Antarctica.

Fertilizers are chemicals added to plants to help them grow.

The neem tree is a tropical tree that is popular in Ayurvedic medicine.

The lotus is a sacred flower in Hinduism and Buddhism, and is also a popular ornamental plant.

The rubber tree is the source of natural rubber.

The sundew plant is another carnivorous plant that traps insects on sticky tentacles.

Apples float on water because they are 25% air.

The Great Barrier Reef is the largest living structure on Earth.

Cranberries also float on water as they have small pockets of air.

Glue used to be made from bluebell flower juice.

Oak trees can live for 1,000 years with an average lifespan of 600 years.

The colour of carrots was originally purple not orange. Its new colour is a result of thousands of years of cross-breeding by humans.

Asian watermeal (Wolffia globosa) is the world's smallest flowering plant, with a diameter of 0.004–0.008 inches.

The oceans contain about 85% of all plant life on Earth.

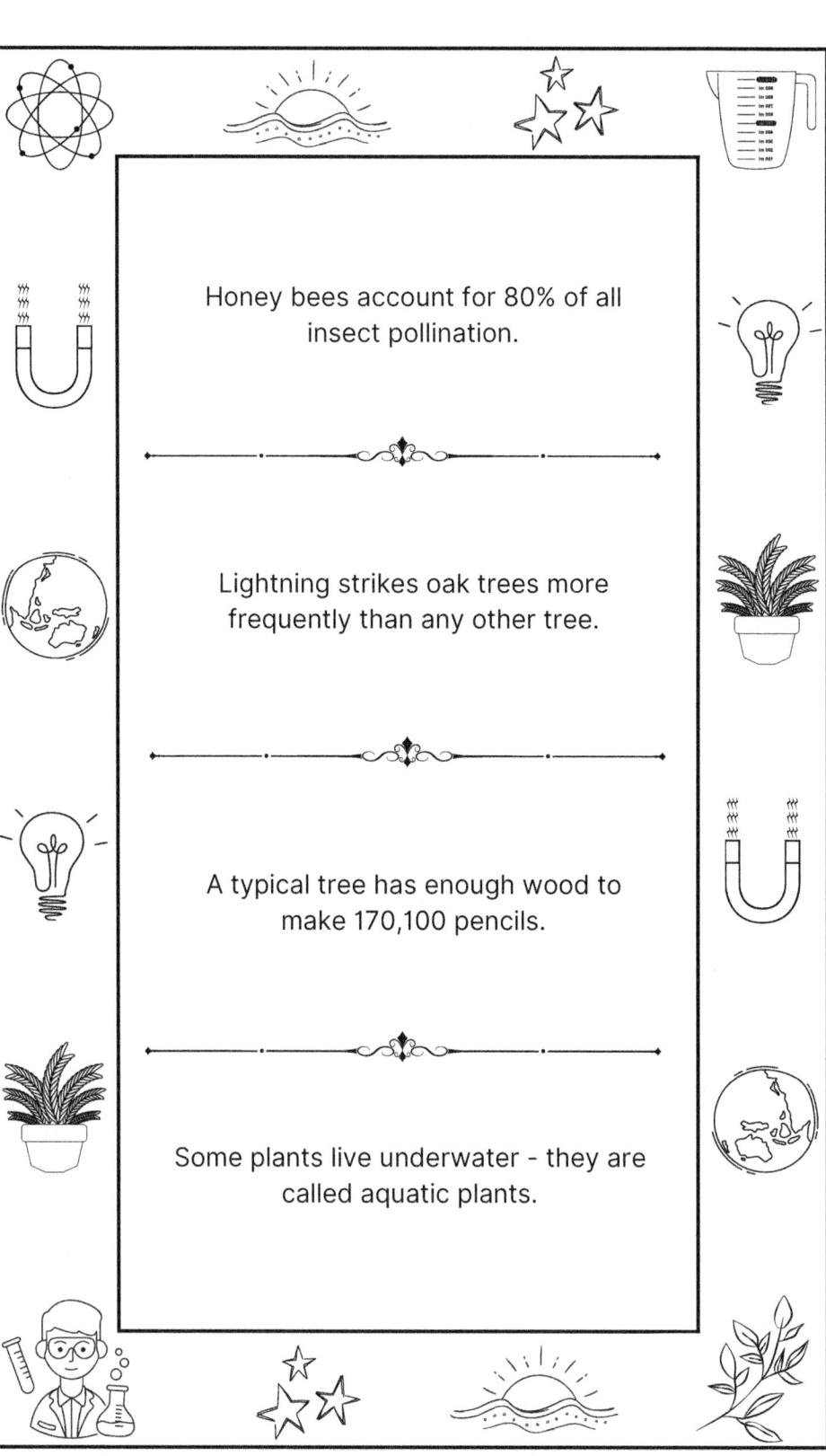

Honey bees account for 80% of all insect pollination.

Lightning strikes oak trees more frequently than any other tree.

A typical tree has enough wood to make 170,100 pencils.

Some plants live underwater - they are called aquatic plants.

The Amazon Rainforest is home to more than two-thirds of the world's population of plant species.

Forests cover around 31% of the Earth's land area

Peanuts are not nuts but are related to beans and lentils.

The average strawberry contains 200 seeds.

Trees have been on our planet for at least the last 370 million years.

Over 25% of all known natural medicines have been discovered in the rainforests.

Bamboo grows on every continent except for Europe and Antarctica.

Over 100 billion bananas are consumed every year.

Famous Scientist

Alexander Graham Bell was an inventor who developed the telephone.

John Logie Baird, a Scottish engineer, who invented the first mechanical television.

Albert Einstein was a German theoretical physicist who is regarded as one of the finest physicists of all time.

Albert Einstein is known for his theory of relativity and famous equation, $E=mc^2$.

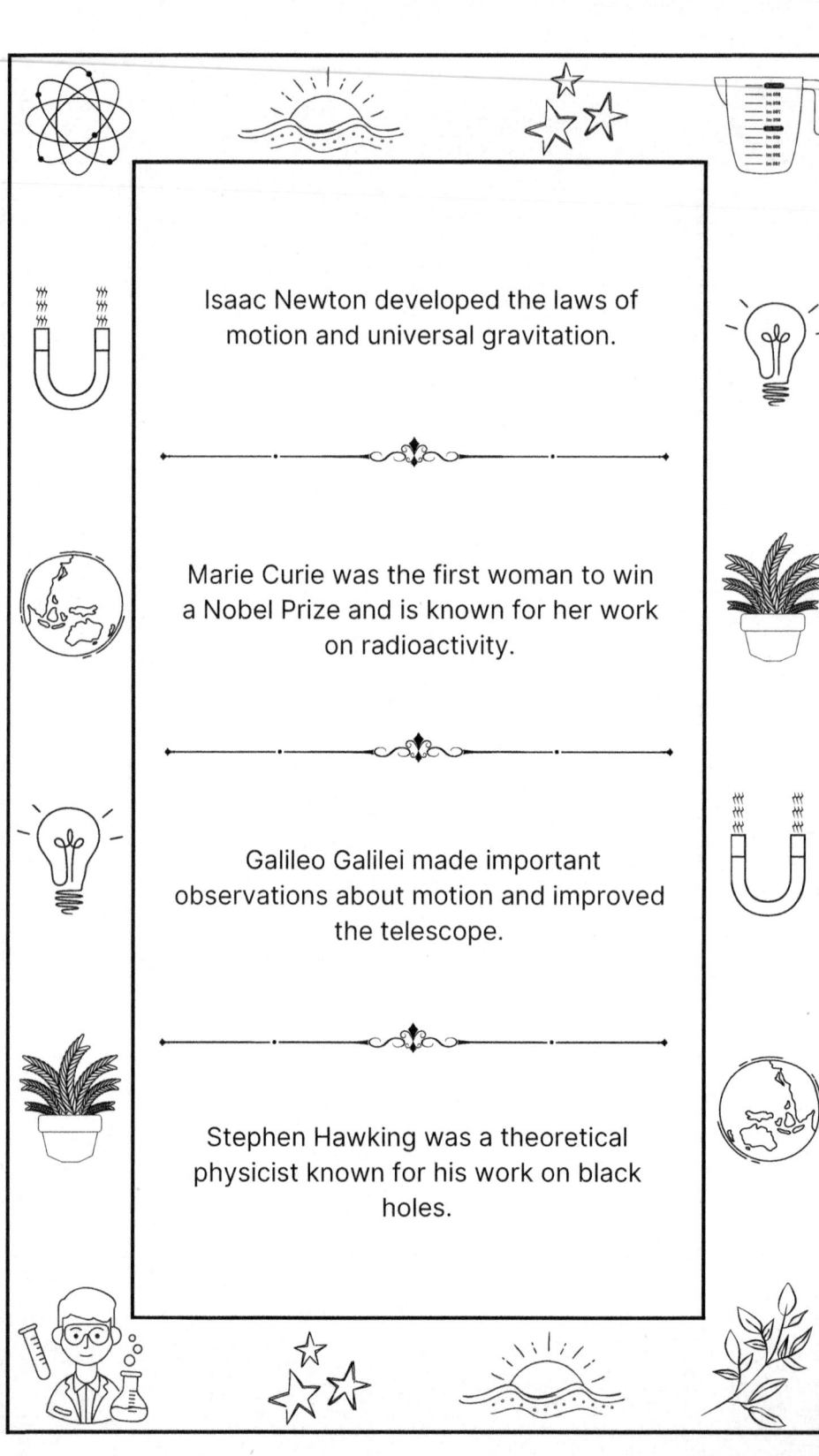

Isaac Newton developed the laws of motion and universal gravitation.

Marie Curie was the first woman to win a Nobel Prize and is known for her work on radioactivity.

Galileo Galilei made important observations about motion and improved the telescope.

Stephen Hawking was a theoretical physicist known for his work on black holes.

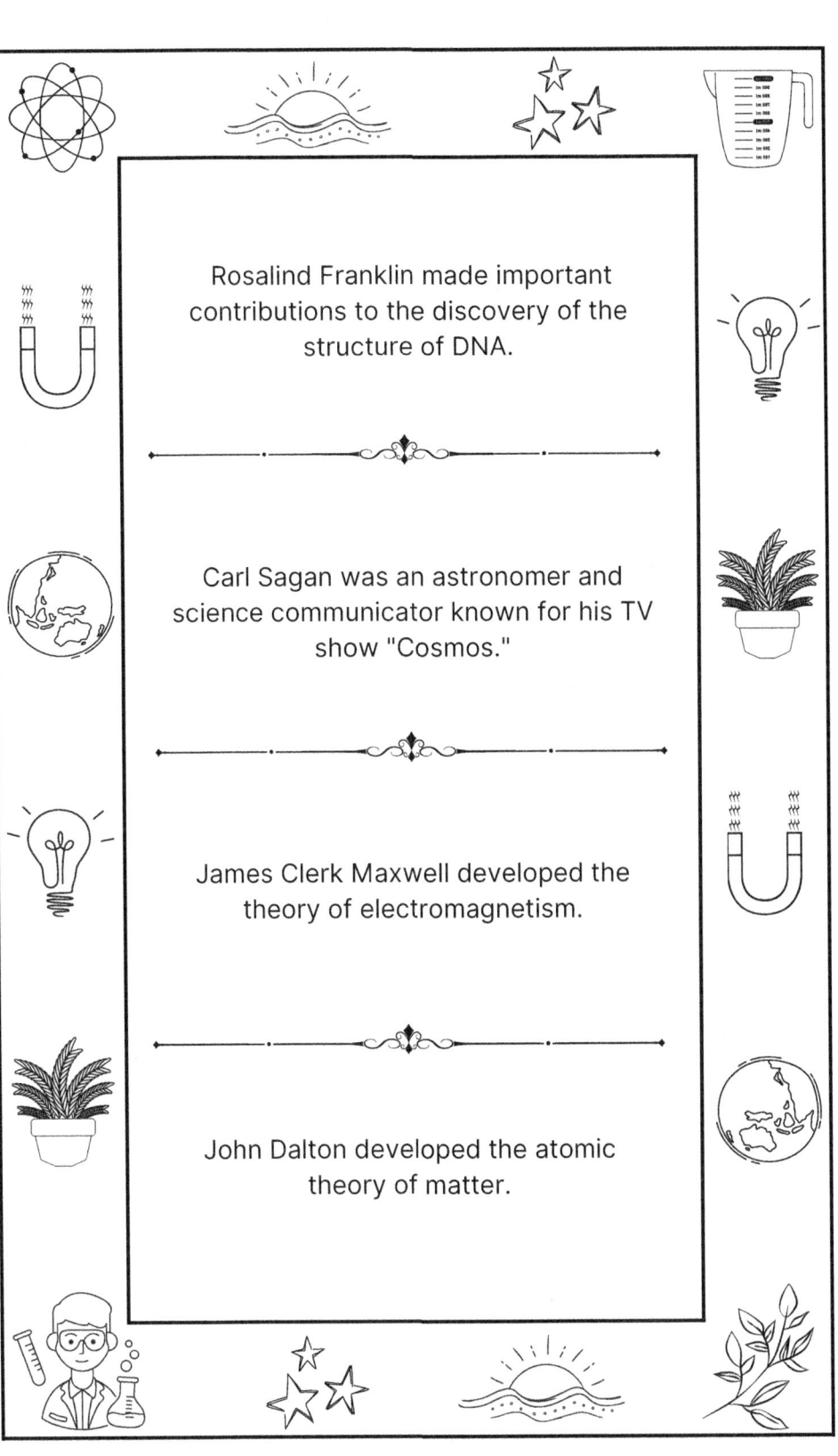

Rosalind Franklin made important contributions to the discovery of the structure of DNA.

Carl Sagan was an astronomer and science communicator known for his TV show "Cosmos."

James Clerk Maxwell developed the theory of electromagnetism.

John Dalton developed the atomic theory of matter.

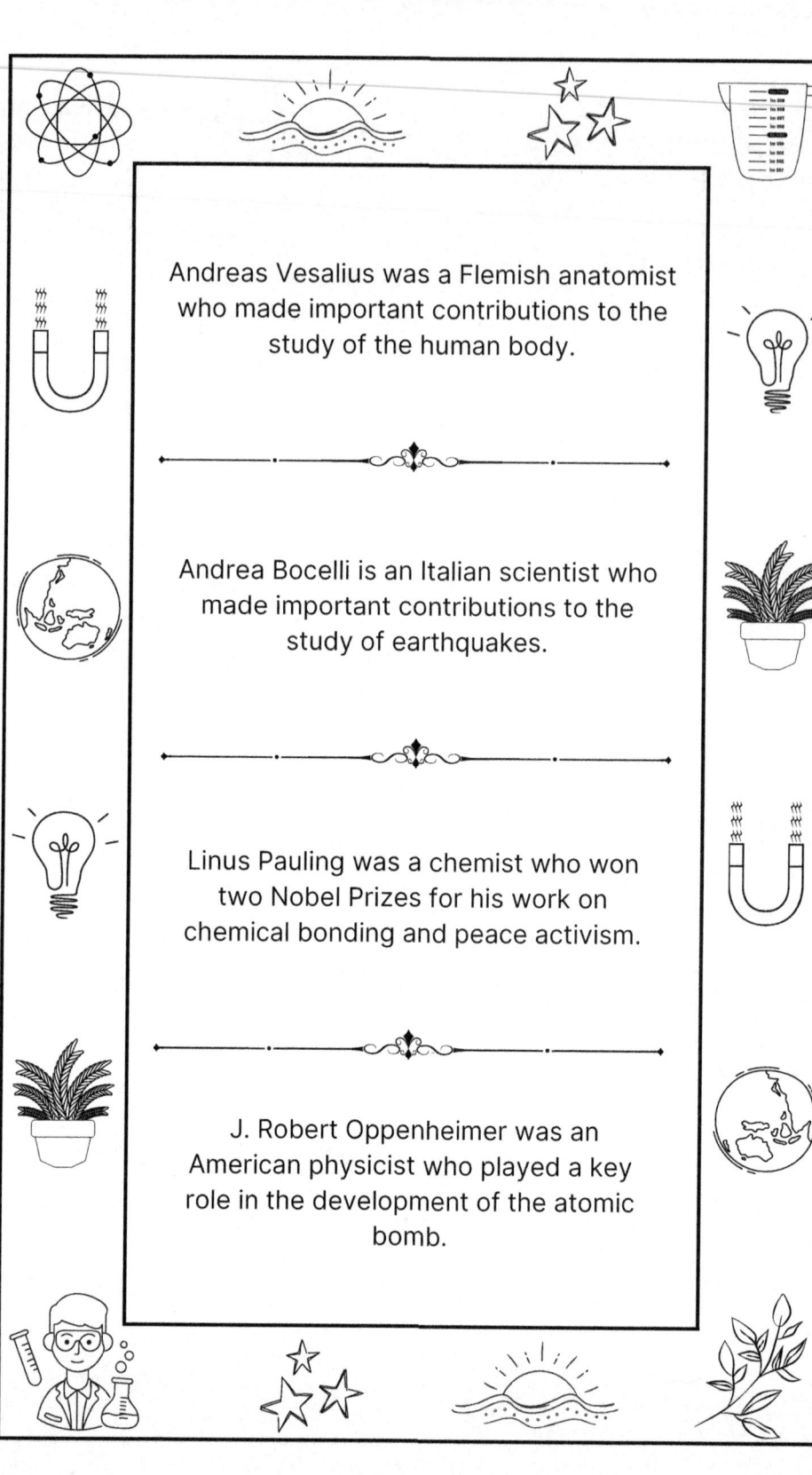

Andreas Vesalius was a Flemish anatomist who made important contributions to the study of the human body.

Andrea Bocelli is an Italian scientist who made important contributions to the study of earthquakes.

Linus Pauling was a chemist who won two Nobel Prizes for his work on chemical bonding and peace activism.

J. Robert Oppenheimer was an American physicist who played a key role in the development of the atomic bomb.

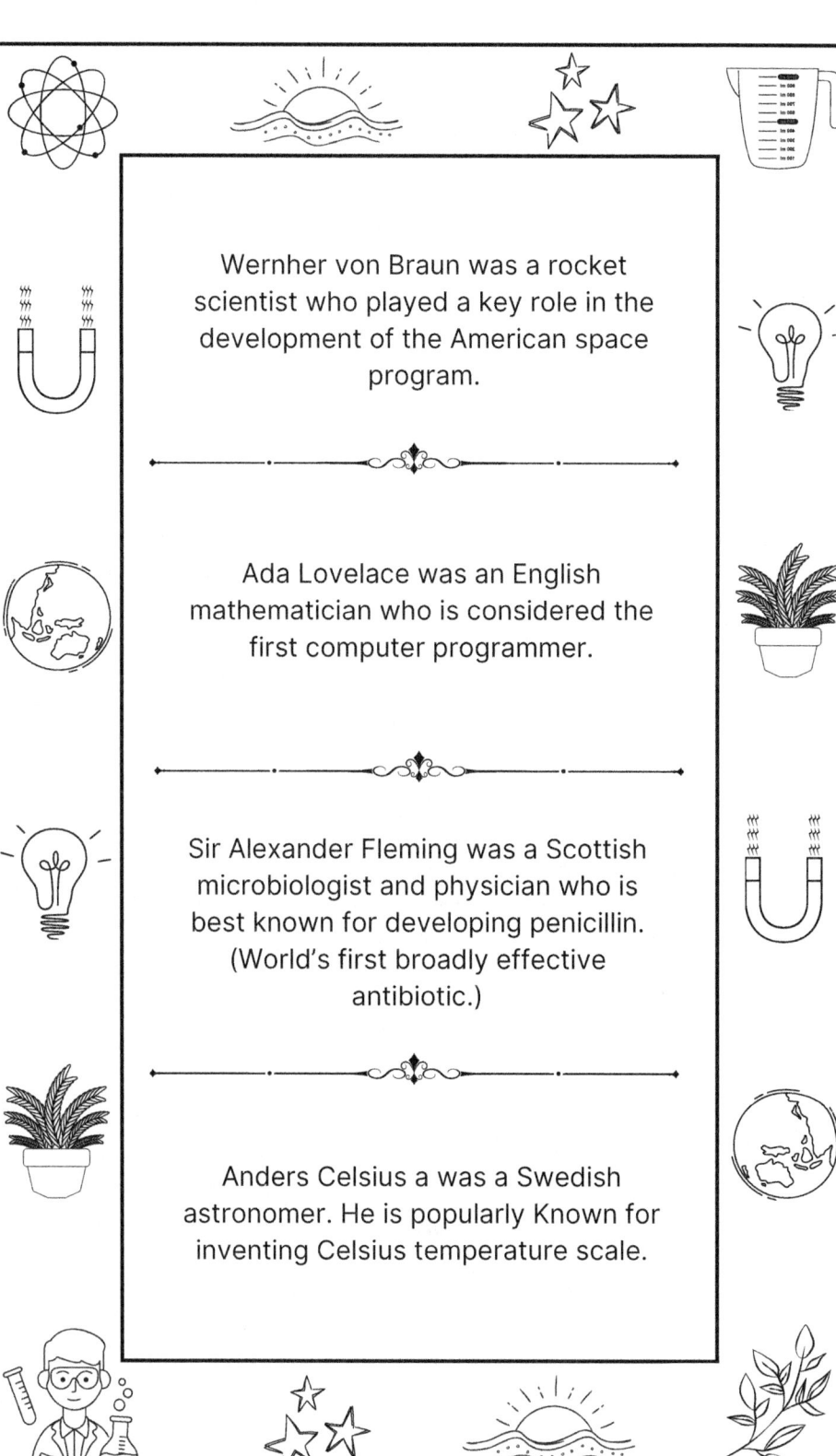

Wernher von Braun was a rocket scientist who played a key role in the development of the American space program.

Ada Lovelace was an English mathematician who is considered the first computer programmer.

Sir Alexander Fleming was a Scottish microbiologist and physician who is best known for developing penicillin. (World's first broadly effective antibiotic.)

Anders Celsius a was a Swedish astronomer. He is popularly Known for inventing Celsius temperature scale.

Gerty Cori was the first woman to win the Nobel Prize in Physiology or Medicine for her work on carbohydrate metabolism.

Hedy Lamarr was an actress who also made important contributions to the field of wireless communication.

Sophie Germain was a French mathematician who made important contributions to the study of elasticity and number theory.

Theodosius Dobzhansky was a Russian-American geneticist who made important contributions to the understanding of evolution.

William Harvey was an English physician who made important contributions to the understanding of the circulatory system.

Samuel Morse was an inventor who developed the Morse code and the electric telegraph.

Rosalyn Sussman Yalow was the second woman to win the Nobel Prize in Physiology or Medicine, for her work on radioimmunoassay.

Max Planck was a German physicist who developed the theory of quantum mechanics.

John Napier was a Scottish mathematician who developed the concept of logarithms.

Antoine Lavoisier is considered the "father of modern chemistry" for his work on the nature of matter.

Jane Goodall is a famous primatologist known for her work on chimpanzees.

Enrico Fermi was an Italian physicist known for his work on nuclear reactions.

Guglielmo Marconi was an Italian inventor who developed the first practical radio.

Thomas Edison was an inventor who developed the light bulb and many other important technologies.

Avicenna was a Persian philosopher and scientist who made significant contributions to medicine.

Francis Crick and James Watson discovered the structure of DNA, winning the Nobel Prize in Medicine.

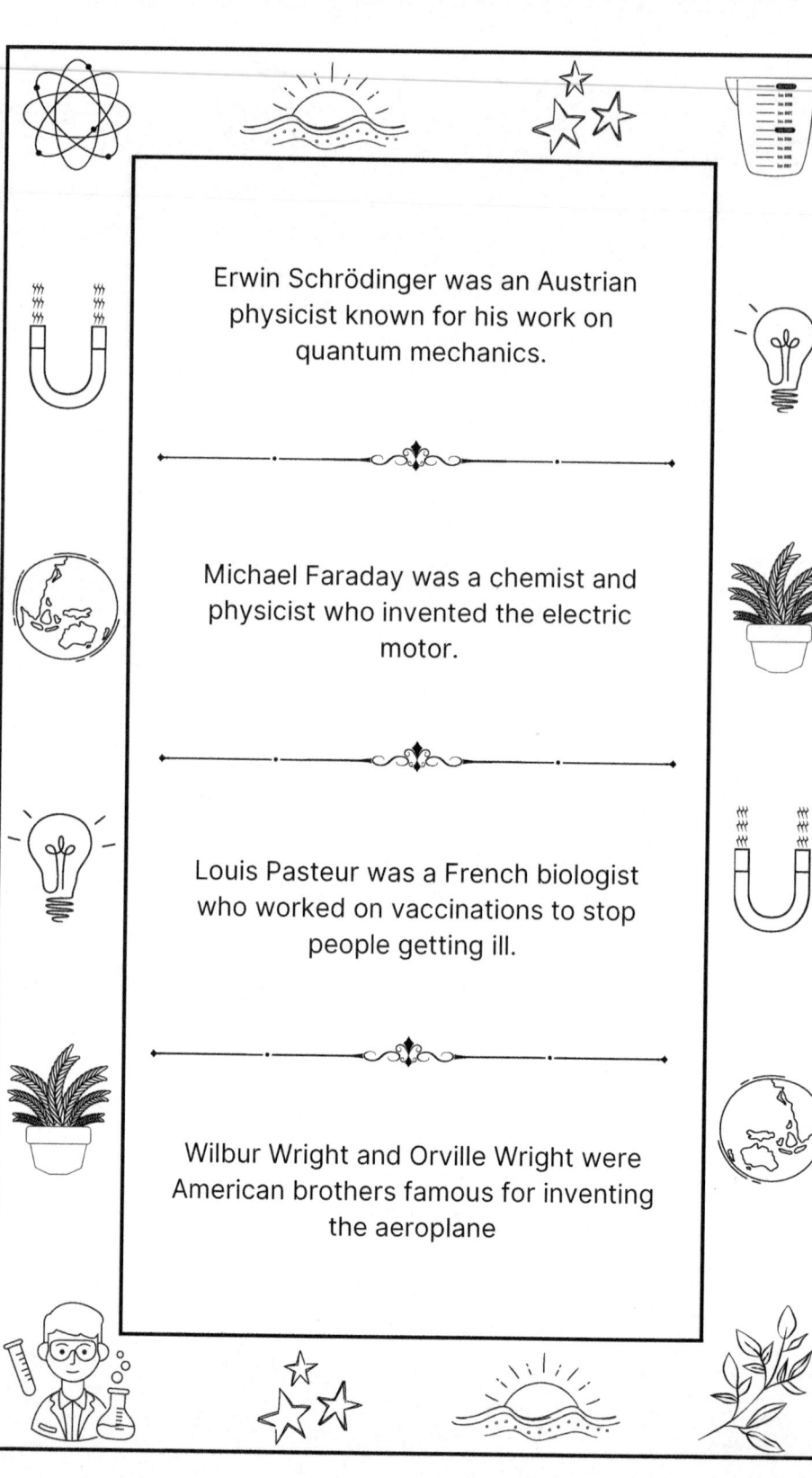

Erwin Schrödinger was an Austrian physicist known for his work on quantum mechanics.

Michael Faraday was a chemist and physicist who invented the electric motor.

Louis Pasteur was a French biologist who worked on vaccinations to stop people getting ill.

Wilbur Wright and Orville Wright were American brothers famous for inventing the aeroplane

Alfred Wegener was a German meteorologist and geophysicist who proposed the theory of continental drift.

Nikola Tesla is a Serbian-American engineer and scientist who made many inventions geared towards the application of electric power.

Georg Simon Ohm was a German physicist who developed the law of electrical resistance.

Selman Waksman was a microbiologist who developed important antibiotics.

Alfred Nobel was a Swedish scientist who invented dynamite and established the Nobel Prizes.

Wilhelm Roentgen was a physicist who discovered X-rays, winning the Nobel Prize in Physics.

Dimitri Mendeleev was a Russian chemist who developed the periodic table of elements

---

Tim Berners-Lee is a computer scientist who developed the World Wide Web.

Printed in Great Britain
by Amazon